00790

W9-BZK-094

NATURAL GASES
IN
MARINE SEDIMENTS

MARINE SCIENCE

Coordinating Editor: Ronald J. Gibbs, Northwestern University

NATURAL GASES
IN
MARINE SEDIMENTS

Edited by
Isaac R. Kaplan

*Department of Geology and
Institute of Geophysics and Planetary Physics
University of California at Los Angeles
Los Angeles, California*

Withdrawn
University of Waterloo

PLENUM PRESS • NEW YORK AND LONDON

Library of Congress Cataloging in Publication Data

Main entry under title:

Natural gases in marine sediments.

(Marine science, v. 3)
"Proceedings of a symposium conducted by the Ocean Science and Technology
Division of the Office of Naval Research . . . held at the University of California
Conference Center, Lake Arrowhead, November, 1972."
Includes bibliographies.
1. Marine sediments—Gas content—Congresses. I. Kaplan, Isaac R., ed. II.
United States. Office of Naval Research. Ocean Science and Technology Division.
GC380.2.G3N37   551.4'6083   74-11492
ISBN 0-306-35503-5

Proceedings of a symposium conducted by the Ocean Science and Technology
Division of the Office of Naval Research on Natural Gases in Marine Sedi-
ments and Their Mode of Distribution held at the University of California
Conference Center, Lake Arrowhead, November 1972

PREFACE

In July 1972, the U.S. Office of Naval Research identified several areas that it interpreted as being of interest to the U.S. Navy. Four of these research areas were then selected for their special importance in understanding physical processes on the ocean floor. In some of these, a great wealth of data has accumulated over the past two or three decades, but controversy exists in the interpretation of the results. In others, new techniques have recently been devised that could lead to the collection and synthesis of new information. There was yet a third area in which little study had been undertaken and the results available appeared of great potential importance. The latter subject constitutes the title of this volume.

To assess the information available and to facilitate plans for further research in the fields of interest that had been identified, the U.S. Office of Naval Research sponsored four symposia. The first was held in November 1972 at the University of California Conference Center, Lake Arrowhead. The title of the symposium was "Natural Gases in Marine Sediments and Their Mode of Distribution". Twenty lectures were presented over a three-day period. All but two participants at this symposium subsequently submitted papers, which are published in this volume. In addition, Dr. K.O. Emery, who did not attend the symposium, supplied a manuscript on a topic most relevant to the subject matter discussed.

I would like to take this opportunity to thank the U.S. Office of Naval Research for funds to invite participants to Lake Arrowhead as well as for the subsequent compilation and editing of the present volume. Particular mention should be made here to Dr. Alex Malahoff, who took a personal interest in ensuring the success of the symposium. I would like to give credit to Mrs. Rosemond Cline for her expert assistance in all phases of the compilation of this volume. She was responsible for the typing, grammatical corrections, reference checking, index compilation and other hidden duties that fall to an associate editor. I acknowledge with thanks her conscientious efforts.

I.R. Kaplan
Editor

Los Angeles, California
2 March 1974

v

CONTENTS

vii

INTRODUCTION

That gases are present in marine sediments has been known for a long time. The nature and distribution of these gases have, however, received little investigation. Early studies were almost exclusively directed toward bay or estuarine environments and were essentially extensions of work undertaken on soil, swamps, and lakes. Much of the reliable work on shallow water marine environments was undertaken by Koyama and associates at Nagoya University [Koyama, 1953]. The first successful attempt to obtain quantitative data in deeper water sediments was made by Emery and Hoggan [1958]. They captured sediment in a core barrel that was capped and taken to the laboratory for gas removal and analyses by mass spectroscopy. From their analytical method, they were able to measure methane and several other volatile homologs. Although their techniques could not yield accurate data, as a first approximation, the results they obtained for shelf sediments off southern California are still valid. Subsequent studies by Reeburgh [1969] and others have produced quantitative data for specific gases in specific environments.

All these studies, of course, were undertaken in continental margin sediments. Little or no information was available for continental rise or abyssal floor sediments, although ZoBell [1947] and his students had reported culturing a variety of gas-producing bacteria from deep sea sediments such as the Gulf of California. The first evidence for the nature of the gas in abyssal sediments came from studies initiated by the Deep Sea Drilling Project under the sponsorship of NSF through JOIDES. Results of these measurements have been summarized by Claypool et al. [1973]. An observation reported by Stoll et al. [1971] on anomalous acoustic reflections in marine sediments on the Blake Outer Ridge, off the east coast of the USA, was interpreted by the authors to have resulted from the in situ presence of gas hydrates within the sediment column. This interpretation drew the attention of marine scientists to a hitherto unexplored research area, with potentially important implications in several disciplines.

The symposium held in November 1972 at the University of California Conference Center, Lake Arrowhead, under the sponsorship of the U.S. Office of Naval Research, was a direct result of the Stoll et al. [1971] publication. The participants gathered at the

1

symposium were specialists in microbiology, chemistry, geology, and physics, a truly cosmopolitan group to discuss and debate a problem of environmental significance.

NATURE, ORIGIN AND CONTENT OF GASES IN MARINE SEDIMENTS

Gas in marine sediment may arise from four sources: atmospheric, from gases originally dissolved in sea water; gases produced during early diagenesis from biogenic degradation of organic matter; gas diffusing upward from depth where it has been produced by the thermocatalytic cracking of more complex organic compounds; and gas produced by submarine volcanic or geothermal processes.

Two reactive atmospheric gases are present in significant quantities in sea water; they are oxygen and carbon dioxide (present in sea water as the bicarbonate ion). Oxygen will react with a large variety of detrital and igneous minerals and biological products of metabolism. It is believed to be rapidly removed in the surface layers of most sediment, either by oxidation of the minerals and organic matter or by respiration. Its rate of removal will depend on the availability of reactive substances and population size and nature of the benthonic biota. Bioturbation can reintroduce oxygen into the surface sediment layers. There is presently no reliable information on the depth distribution of dissolved oxygen in the interstitial water of marine sediments. Atmospheric carbon dioxide is probably a small component of the total dissolved bicarbonate in most marine sediments.

The most important mechanism for generating gases in marine sediment is the microbiological degradation of organic matter. As discussed by *Mechalas* [pp. 11-25] and by *Claypool and Kaplan* [99-139] in this volume, it is possible to suggest a series of reactions down a sediment column, beginning with oxygen consumption at the top, which results in CO_2 production. Beneath this zone, the small amount of nitrate present in sea water is rapidly denitrified and molecular N_2 will be produced. The sediments are now anoxic, and fermentative degradation of organic matter begins with the consequent release of carbon dioxide and hydrogen. Hydrogen gas does not accumulate and is apparently rapidly used for the microbiological reduction of dissolved sulfate to produce hydrogen sulfide (H_2S). This gas is highly corrosive and will react relatively quickly with minerals to form iron sulfides. It is therefore only present in organic-rich rapidly-depositing sediment near the sea water sediment interface [*Goldhaber and Kaplan, 1974*] where it is continually being produced. When all the sulfate has been reduced, the hydrogen is used by another group of bacteria to reduce carbon dioxide to methane (CH_4). The exact nature of these organisms is not known, although *ZoBell* [*1947*] attempted to isolate methanogenic strains. Because H_2S and CO_2 can react and be removed as insoluble precipitates, or else dissociate into ionic forms, CH_4 and N_2 become the dominant gases in marine sediment.

Thermocatalytic degradation of organic matter first results in CO_2 production by decarobxylation at temperatures as low as 65°C in a time period of seven days (unpublished results). In experiments undertaken in the author's laboratory, only a trace of methane was produced in heating a recent marine sediment to 165°C for 74 days, the main gaseous component being CO_2. *Hoering and Abelson* [*1963*], however, were able to generate methane from a Green River Shale by heating in the temperature range of 185-230°C. As the temperature of heating was increased, the ratio of methane/ethane and higher homologs ($C_1/C_2 + C_3 +...$) decreased [*Abelson, 1963*]. In petroleum-rich regions (i.e., the Texas Gulf Coast), the ethane content increases with increasing depth of burial and may reach 2 per cent of the methane content [*Buckley et al. 1958*]. However, in normal marine sediments, the ethane content is generally 3 to 6 orders of magnitude less than that of methane. Results presented by *Claypool et al.* [*1973*] in deep sediments obtained from the Deep Sea Drilling Program (DSDP) indicate that when ethane is present, it increases with depth relative to methane content and its relative concentration is higher in areas of high heat flow. With the one exception of Leg 10, Hole 88 of the DSDP, drilled in the Gulf of Mexico, where $C_2H_6/CH_4 \simeq$ 3-8 per cent, all other sites yielded a ratio < 0.1 per cent. Hence, ethane is of little significance in the total gas content of marine sediments.

The fourth major source of gas in marine sediments is probably submarine volcanism. The major gas produced in geothermal wells is carbon dioxide [*Lyon,* this volume, pp. 141-150]. However, depending on the depth of the igneous source and its relationship to the continental margin, a variety of other gases, such as CH_4, H_2, N_2, H_2S, SO_2, and halogens may also be produced. Such gases have not yet been sampled in marine sediments, although gas bubbles have been observed to rise in water off the northeast coast of New Zealand in a volcanically-active region [*Duncan and Pantin, 1969*].

The above discussion should serve to indicate that methane, carbon dioxide, and nitrogen are the three major gases resulting from the various synthetic and degradative processes. Because a successful collecting technique has not yet been described, which can capture and maintain an intact sample of gas from the sediment, a precise measure of gas volumes has not yet been made. *Reeburgh* [this volume, pp. 27-45] measured dissolved CH_4 volumes in shallow (100 cm) cores from Chesapeake Bay to be in the range of 85 to 150 ml/l interstitial water, N_2 content was $\simeq$ 10 ml/l at the sediment surface and decreased with depth to 1-2 ml/l. This was explained by Reeburgh as resulting from upward diffusion of methane and nitrogen scavenging.

Whelan [this volume, pp. 47-61] found methane contents in shallow coastal marsh sediments to be only 0.13-7.4 ml/l interstitial water. It is probable that in this environment, methane generation may in part be inhibited by the presence of sulfate (perhaps flushed in periodically) and also methane loss may occur out of the sediment by diffusion.

McIver [this volume, pp. 63-69], using gas volume measurements in fixed volume cans containing known weights of sediment, measured a concentration range for methane from a few tens of parts per million to 215,000 ppm by volume in one sample. Such a high content compares very favorably with gas contents of organic-rich shales.

Hammond [this volume, pp. 71-89] found that gas pockets in cores from Cariaco Trench contain about 90 per cent CH_4 and 8 per cent CO_2. By use of volume and pressure corrections, he was able to calculate the content of CH_4, N_2 and C_2H_6 to be 440 ml/l, 1.4 ml/l, and 90 µl/l, respectively.

Claypool and Kaplan [this volume, pp. 99-139] used an indirect method for calculating the volume of biogenic methane generated in the sediment. They assumed that most of the methane is formed within the sediment column by the reduction of CO_2 with hydrogen, i.e., $CO_2 + 4H_2 \rightarrow CH_4 + 2H_2O$. During such a process, there will be an isotopic enrichment in C^{12} in the CH_4 formed, whereas the remaining CO_2 will be enriched in C^{13}. The amount of methane formed can be calculated knowing: the amount of total dissolved CO_2, the isotopic ratios of CO_2 and CH_4, and the instantaneous fractionation factor α, which is the measure of isotopic enrichment in C^{12} in methane (relative to CO_2) when first formed (in this case $\alpha = 1.07$). Applying a modified Rayleigh distillation equation to the data, Claypool and Kaplan calculated that a minimum of 450 ml CH_4/kg interstitial water forms in organic-rich sediment by the method suggested above. The amount will increase if it is found that CH_4 can form from carboxylic acids (i.e., acetic acid) present in the sediment.

PHYSICAL STATE OF GASES IN SEDIMENT

It has been generally assumed that gases in sediment are either dissolved in the interstitial water or exist as free gas bubbles. However, the interpretation expounded by *Stoll et al.* [*1971*], is that gases may also exist in a solid form as gas hydrates. Knowledge of solid gas hydrates has been available since 1811, when Humphrey Davy first published his experimental data for chlorine gas. In the intervening years, a great deal of investigation has been undertaken on the formation processes, crystallography, stoichiometry, host-guest relationships, and more recently, their distribution in nature as compared to the laboratory. Some of these aspects have been discussed by *Claypool and Kaplan* [pp. 99-139], *Miller* [pp. 151-177], *Hand et al.* [pp. 179-194] and *Hitchon* [pp. 195-225] in this volume, and they are briefly summarized below.

Gas hydrates are a special form of clathrate compounds in which crystal lattice cages or chambers consisting of host molecules enclose guest molecules. In the case of the gas hydrates, the host molecules are water and the guest molecules may be a large variety of gases including argon, nitrogen, carbon dioxide, hydrogen sulfide, methane, ethane, halogens, and numerous other small molecules. Gas hydrates are solid compounds, resembling ice or wet snow in appearance. They crystallize in two structures. Structure (I) contains 46 molecules in the unit cell, which includes 2 pentagonal dodecahedra (12-sided polyhedra) and 6 tetrakaidecahedra (14-sided polyhedra). Gas molecules smaller than 5.1 Å fit into the pentagonal dodecahedra, whereas molecules 5.8 Å or less in diameter can fit into the tetrakaidecahedra. Thus, there are 8 cavities and 46 host (water) molecules. In the case of a simple gas hydrate, such as methane or carbon dioxide, the relationship $CH_4/H_2O = (1)/(5-3/4)$. It is also possible to obtain mixed hydrates composed of more than one gas (i.e., CH_4; $CO_2 \cdot 5\text{-}3/4 H_2O$). The stoichiometric relationship represented above is, however, apparently ideal as guest cages may not always be filled. Structure (II) contains 136 molecules of water in the form of 16 pentagonal dodecahedra and 8 hexakaidecahedra (16-sided polyhedra). Molecules with diameters of 6.7 Å or less will fit into the hexakaidecahedra. These structures tend to form in the presence of large guest molecules (i.e., cyclopropane or chloroform) and are probably not important in most natural environments.

The conditions governing the formation of gas hydrates follow the Gibbs phase rule, where

Degrees of freedom = Number of components -

number of phases + 2

Five possible equilibrium phases can form in a gas-water mixture. These are:

1) Gas saturated with water vapor,
2) Water saturated with gas,
3) Liquid gas saturated with water,
4) Ice,
5) Hydrate.

The conditions controlling phase equilibria for a large number of gas-water systems have been established experimentally and the thermodynamic properties have been described by application of statistical thermodynamics [see *Miller*, pp. 151-177 this volume for details].

In general, hydrates form at high pressure and low temperature, if the gas is supersaturated under the conditions specified. As is the case in formation of ice from water, salts lower the temperature of hydrate formation. In gases present in marine sediment, the relative ease for hydrate formation (in terms of P and T) is $H_2S > C_2H_6 > CO_2 > CH_4 > N_2$. Or, for the same sequence of gas, the dissociation pressure at 0°C for the hydrates is 0.92, 5.2, 12.5, 26, and 160 atm. Hence, assuming supersaturation of gas and bottom temperatures near zero, methane hydrate could form beneath a water column of 260 meters, and along continental shelves or shallow seas in temperature regions where bottom temperatures may be typically about 5°C, methane hydrates would be stable under a 500 meter column of water.

The stability of the hydrate within the sediment is in part controlled by hydrostatic pressure of the overlying water and in part by the geothermal gradient within the sediment. Hence, although there is a linear pressure increase with depth of burial, the temperature increase is relatively more important, so that at some depth in each sediment column the hydrate will become unstable depending on the depth of water column and geothermal gradient in the sediment. This has been shown diagrammatically by *Claypool and Kaplan* [Figure 15, p. 128], *Hand et al.* [Figure 9, p. 192], *Stoll* [Figure 3, p. 239], and *Bryan* [Figure 4, p. 304] in this volume. For example, where bottom temperatures are 2°C and the depth of water column is 2 km, Claypool and Kaplan (Figure 15) showed that methane hydrate will be present to a depth of ≃ 1000 meters in the sediment, assuming a geothermal gradient of 0.035°C/m.

Investigations of gas hydrates were accelerated after 1935, when it was found that gas hydrates often clogged gas lines if moisture was present. This practical problem was overcome by drying the gas. Experimental conditions for formation and decomposition of gas hydrates, including equipment used, are described in this volume by *Baker* [pp. 227-234] and *Stoll* [pp. 235-248].

ACOUSTIC PROPERTIES OF GAS-RICH SEDIMENT

The acoustic properties of bubble-containing water has received a great deal of attention, and is briefly reviewed by *Hampton and Anderson* [this volume, pp. 249-273]. The basic principle controlling sound velocity is the fact that bubbles in water are capable of vibratory motion with a sharply peaked resonance at the fundamental pulsation frequency. The gas bubble acts as a cavity and the surrounding water as the vibrating mass of an acoustical oscillator. Maximum attenuation of sound transmitted through a screen of bubbles occurs at wave frequencies near resonance, but decreases exponentially at frequencies distant from resonance.

Few studies have been directed to the attenuation of sound propagation in bubble-containing sediments. Much of the investigations that were conducted have been undertaken by Hampton and co-workers at Applied Research Laboratories (ARL), Texas. In general, it is believed that gas-containing sediment results in a marked attenuation in sound velocity, although some workers claim that a velocity increase can result in gas-containing sediment [see *Hampton and Anderson,* this volume, p. 259]. Some of the controversy may in part be due to the difficulty in determining whether bubbles in fact do exist in the sediment. As methane does not begin to form until sulfate has been reduced, it may not become sufficiently saturated to form bubbles in marine sediment until a depth of several meters has been reached. Equipment for *in situ* measurement of sound transmission in sediments has been built at ARL and the results demonstrate the attenuation behavior of gas-rich sediment. This laboratory has also built a gas collection sampler to quantitatively determine the gas content of surface sediment [see *Hampton and Anderson,* this volume, p. 263].

To determine why sediments show high acoustical attenuations (sometimes referred to as 'acoustically-turbid sediment'), *Schubel* [this volume, pp. 275-298] collected cores from Chesapeake Bay sediments. Three techniques were then applied to obtain a measure of the gas and/or bubble content. First, gas was driven from the cores and collected. Second, cores were frozen immediately after collection and later examined by X-ray radiography for bubble tracks. Third, sediment was compressed and the compressibility compared with overlying water.

Schubel found that in acoustically-turbid coarse sediment, gas was rapidly removed. However, this was not so in fine-grained sediment. On the other hand, X-radiographs showed distinct bubble locations in 'acoustically-turbid' fine sediment. Similarly, bubble-containing sediment had a compressibility approximately 2 orders of magnitude greater than that of water, whereas the compressibility of sediment in acoustically-clear areas could not be measured because it was below the sensitivity of the apparatus used.

Sound velocity in gas hydrate is greater than in water, and this in part may have given rise to the controversy relating to increase or decrease of sound velocity in gas-rich sediments. Reliable data have been obtained by *Stoll* [this volume, pp. 235-248] in laboratory experiments. His results give strong support to the initial report of gas hydrates existing in sediments [*Stoll et al. 1971*]. In that report, the authors showed that for DSDP Leg 11, Holes 102 and 104, on the southwest and northwest flanks of the Blake-Bahama outer ridge, the sound velocity from near the sediment surface to a depth of 620 meters was in the range of 2.0-2.2 km/sec. The normal velocity for near surface unconsolidated sediment is 1.5-1.6 km/sec, which increases nearly linearly with depth due to compaction and water loss. Furthermore, as explained in detail by $Bryan$ [this volume, pp. 299-308], below 620 meters for cores 102 and 104, a sharp transition occurred over a 20 meter depth in which the sound velocity dropped to 1.75 km/sec and then increased again over a range of 100-200 meters. Sample retrieval did not indicate any unusual lithology that may have been responsible for the acoustic anomaly; hence, it was assumed to be an ephemeral component that decomposed. This added further support to the concept of a gas hydrate as the responsible component.

The drop in velocity between 620-640 meters was interpreted as resulting from a layer of gas bubbles forming an acoustically-turbid reflector. This explanation appears satisfactory, but for the fact that during drilling of the holes, resistance increased in reaching the 620 meter interface rather than decreased.

One possible explanation for such behavior comes from the studies of $Claypool$ and $Kaplan$ [this volume, pp. 99-139]. They suggest that during biological methane formation, calcium or other metal (Fe) carbonates may form, and may therefore act as a cement. This occurs because the bacteria reduce carbon dioxide to methane, whereas at the pH of interstitial water the dissolved carbon dioxide species are present as bicarbonate. Hence, in order to maintain electrical neutrality, an equivalent amount of carbonate will form and can precipitate on supersaturation.

This suggestion is illustrated below

$$2HCO_3^- \begin{cases} H_2CO_3 \rightarrow CO_2 + H_2O \xrightarrow{4H_2} CH_4 + 3H_2O \\ CO_3^= + Me^{++} \rightarrow MeCO_3 \end{cases}$$

It is quite probable that such gas hydrates are more common in deep water organic-rich sediment than had been previously suspected. They should be located mostly along the continental margins where organic matter is available and the water column height is sufficient to stabilize the hydrate. $Emery$ [this volume, pp. 309-317] suggests that alternating acoustically dark and light triangular features detected at depths of 2000-5000 meters off western Africa display internal structure and acoustic properties that may be due to local centers of cementation by gas hydrates. These structures have been referred to as 'pagoda structures', and according to

Emery are features that may be common in fine-grained sediments of continental rises and abyssal plains of the world oceans.

Miller [this volume, pp. 151-177] has speculated that gas hydrates are present on Mars and possibly the outer planets (and perhaps comets), based on their known characteristics. He has also suggested that air trapped in Antarctic ice may combine with water under conditions of low temperature and deep burial to form clathrates. Russian scientists and engineers have demonstrated the presence of hydrocarbon-hydrates in the Vilyuy basin and the northern portion of the West Siberian basin. These are respectively the 19th and 20th largest known gas pools in the world. According to *Hitchon* [this volume, pp. 195-225], potentially commercial occurrences of natural gas hydrates are restricted to sedimentary basins with extensive areas of relatively thick, continuous permafrost. Such areas are known to exist in the Arctic slope petroleum province of Alaska, the Mackenzie Delta, and Arctic Archipelago of Canada, beside those known in the USSR. The possibility that large volumes of methane may be trapped as gas hydrates in continental margin sediments should be examined and its potential evaluated.

REFERENCES

Abelson, P. H., Organic geochemistry and the formation of petroleum, *Sixth World Petrol. Cong. Proc.*, Frankfurt/Main, Sec. I, pp. 397-407, 1963.

Buckley, S. E., C. R. Hocott, and M. S. Taggart, Jr., Distribution of dissolved hydrocarbons in subsurface waters, in *Habitat of Oil*, edited by L. G. Weeks, pp. 850-882, Amer. Ass. Petrol. Geol., Tulsa, Okla., 1958.

Claypool, G. E., B. J. Presley, and I. R. Kaplan, in *Initial Reports of the Deep Sea Drilling Project*, vol. 19, pp. 879-884, U. S. Government Printing Office, Washington, 1973.

Duncan, A. R., and H. M. Pantin, Evidence for submarine geothermal activity in the Bay of Plenty, *N. Z. J. Mar. Freshwater Res.*, *3*, 602, 1969.

Emery, K. O., and D. Hoggan, Gases in marine sediments, *Amer. Ass. Petrol. Geol. Bull.*, *42*, 2174, 1958.

Goldhaber, M. B., and I. R. Kaplan, The sulfur cycle, in *The Sea*, vol. 5, edited by E. Goldberg, pp. 394-454, Wiley Interscience, New York, 1973.

Hoering, T. C., and P. H. Abelson, Hydrocarbons from kerogen, Annual Rept., Director Geophys. Lab., *Carnegie Inst. Wash. Yr. Bk.*, *62*, 229, 1963.

Koyama, T., Measurement and analysis of gases in sediments, *J. Earth Sci., Nagoya Univ., 1,* 107, 1953.

Reeburgh, W. S., Observations of gases in Chesapeake Bay sediments, *Limnol. Oceanogr., 14,* 368, 1969.

Stoll, R., J. Ewing, and G. Bryan, Anomalous wave velocities in sediments containing gas hydrates, *J. Geophys. Res., 76,* 2090, 1971.

ZoBell, C. E., Microbial transformation of molecular hydrogen in marine sediments, with particular reference to petroleum, *Amer. Ass. Petrol. Geol. Bull., 31,* 1709, 1947.

PATHWAYS AND ENVIRONMENTAL REQUIREMENTS FOR BIOGENIC GAS PRODUCTION IN THE OCEAN

Byron J. Mechalas*

Environmental Geology Program
University of Southern California
Los Angeles, California 90007

ABSTRACT

Gases identified in marine sediments and interstitial waters include O_2, CO_2, N_2, NH_3, H_2S and CH_4. The major portion of these gases appears to be the result of coupled oxidation-reduction reactions carried out by microorganisms oxidizing organic matter. Several groups of bacteria are known to be capable of utilizing specific organic or inorganic molecules as the hydrogen acceptors. The environmental conditions under which these reactions take place determine the types of compounds present and the amount of free energy made available to the organisms.

Microbial processes that result in the formation of the gases mentioned above have long been of interest to microbiologists. The apparent uniqueness of the organisms involved has resulted in a great deal of research to determine the chemical pathways and types of transformations that occur. There has also been a very strong applied aspect to some of this research. For example, many of these processes are also important in the fields of plant and soil science, limnology, waste treatment, and several other related fields of endeavor.

Information that has been developed to explain biogenic gas production in the terrestrial sphere can also be related to observations reported from the marine environment. The role of organic carbon in denitrification, the appearance of ammonia under anaerobic conditions, and the formation of small organic molecules essential for methane production can be discussed in the light of known pathways and environmental requirements.

---

*Present address Southern California Edison Co., Rosemead, California 90053.

11

INTRODUCTION

The concentration of various gases dissolved in the oceans is a result of complex interactions between chemical, physical, and biological processes. These interactions may be represented by dynamic kinetic systems involved with gas production and chemical reactions, or passive equilibrium processes dependent on partitioning of the gases between the atmosphere and the water. When the physical parameters (temperature, salinity, etc.) are known, predictions can be made on the amounts of a gas that should be present in solution.

It is not unusual that in given areas of the sea anomalous concentrations of gases, either as excesses or deficiencies over predicted values, are related to a biological action. Living organisms, and especially microorganisms, carry out metabolic processes that can affect the amounts of O_2, CO_2, N_2, H_2, CH_4, H_2S, and other volatile products in the water.

Bacteria obtain energy for growth and cell maintenance from a series of dehydrogenation, or coupled oxidation-reduction reactions. The uniqueness of each microorganism, as far as physiological groupings are concerned, is based upon the type of molecule that can be used as the terminal hydrogen or electron acceptor. This final oxidant is required to maintain the sequence of coenzyme reoxidations which provide for a controlled release of energy from the substrate molecules.

The process of removing an oxidized molecule from solution and replacing it with one that is reduced, alters the chemistry of the surrounding water and imposes a selective pressure on the indigenous microflora, permitting only certain physiological types to survive successfully in the new environment.

GASES AND THE MARINE ENVIRONMENT

Oxidative Processes

The best known terminal acceptor is of course oxygen. During aerobic assimilation, oxygen is rapidly utilized by mocroorganisms, carbon is converted to CO_2, sulfur to $SO_4^=$, nitrogen to NO_3^-, and phosphorus to $PO_4^\equiv$, until energy is no longer available from the molecule for support of the biological process. As the dissolved oxygen concentration in sea water at equilibrium with the atmosphere at 15°C is only about 8 ppm, a point may be reached wherein the demand for oxygen exceeds the rate at which the gas can be introduced from the atmosphere. This may result in the establishment of oxygen depleted zones in various areas of the sea. If living processes are to continue to exist in these anaerobic environments, alternative, and consequently more reduced electron acceptors, would have to be found. The maximum amount of energy is obtained when O_2 is used, with the energy availability then decreasing according to the electron acceptor utilized, as illustrated in Table 1 [*Goldhaber and Kaplan, 1974*].

In nature, microorganisms seldom occur as single species. Usually there are many different metabolic types present, both aerobes and anaerobes, and each group occupies a special ecological niche dependent on substrate availability and restrictions of the physical environment. The species that are best able to adapt to the new conditions are then favored and become dominant. This activity results in a succession of metabolic types that is manifested as either a temporal or spatial progression.

Nitrogen Gas and Nitrates

As indicated in Table 1, once the oxygen supply is exhausted, the next best source of reducing power is nitrate. This ion is the end product of the oxidation of organic nitrogen compounds. The nitrification process proceeding from ammonia to nitrate is a specialized two-step reaction series carried out by a unique group of bacteria as demonstrated in the equations

$$(\text{Nitrosomonas}) \quad NH_4^+ + 1.5O_2 \rightarrow NO_2^- + 2H^+ + H_2O \qquad (1)$$

$$(\text{Nitrobacter}) \quad NO_2^- + 0.5O_2 \rightarrow NO_2^- \qquad (2)$$

TABLE 1. Free Energy ($\Delta F°$) of Reaction between Hydrogen (Equal to Constant Organic Source) and Different Oxidizing Agents*

Reaction	$\Delta F°$ Kcal/Mole
$3O_2 + 6H_2 \rightarrow 6H_2O$	-340.2
$2NO_3^-{}_{(aq)} + 6H_2 \rightarrow 6H_2O + N_2$	-287.4
$1\frac{1}{2}SO_4^={}_{(aq)} + 6H_2 \rightarrow 4\frac{1}{2}H_2O + 1\frac{1}{2}HS^-{}_{(aq)} + 1\frac{1}{2}OH^-$	- 42.1
$1\frac{1}{2}HCO_3^-{}_{(aq)} + 6H_2 \rightarrow 3H_2O + 1\frac{1}{2}CH_4 + 1\frac{1}{2}OH^-$	- 34.0

*From *Goldhaber and Kaplan* [*1973*].

Several groups of bacteria share the ability to utilize this nitrate for coenzyme oxidation. In these reactions the nitrate serves as a direct O_2 substitute with the organic matter going to CO_2 and H_2O and the NO_3^- being converted to N_2. This process, termed denitrification, can account for an increase in the concentration of N_2 over that expected from atmospheric equilibrium in a given water body.

$$4NO_3^- + 4H^+ + 5CH_2O \rightarrow 5CO_2 + 2N_2 + 7H_2O \tag{3}$$

There is a temporal relationship between oxygen, nitrates, and organic matter that can be summarized in the following three statements:

1. When organic matter enters an aqueous environment, dissolved oxygen is consumed by the indigenous microflora, and NO_3^- is produced along with CO_2, H_2O, and other oxidized elements;

2. If the oxygen demand for metabolizing the organic matter exceeds the rate of oxygen replenishment, the environment becomes anaerobic;

3. Once the oxygen is gone, but with organic material still present, NO_3 is utilized and N_2 is released.

The third statement stipulates that organic matter must be present for denitrification to take place. Carbon compounds are the most readily oxidizable energy sources, and are rapidly utilized in the presence of adequate oxygen. Organic nitrogen, on the other hand, is not extensively oxidized until most of the organic carbon has been metabolized. Once the oxidative process has gone to the point of nitrate formation, the available carbon would have been removed, and there would not now be an electron donor available to support the denitrification process. However, denitrification may occur in areas that appear to be deficient in carbon.

In the oceans nitrate concentrations are frequently observed to decrease with depth with a simultaneous increase in the amount of N_2 dissolved in the water [*Richards and Benson, 1961*]. It is possible that when sufficient food material is present, aerobic activity is so intense that anaerobic pockets may form and denitrification can occur. In deep deoxygenated waters, traces of particulate organic material sinking through the water column could sustain very slow denitrification rates. The formation of N_2 may be in equilibrium with the rate of introduction of nitrate which results in zones of continuous denitrification.

An intriguing possibility, and one related to the case

postulated above, is that bacteria may be able to utilize internal stored food reserves for energy to maintain an endogenous respiration through denitrification. This latter possibility was tested in the laboratory.

A large denitrifying bacterial population was developed in a continuous culture device receiving dilute organic sewage as a feed. Once the population had reached an equilibrium condition, wherein all of the incoming organic matter was oxidized and nitrate was the only form of nitrogen present, the air supply to the culture vessel was turned off. The levels of dissolved oxygen and nitrate were then carefully monitored over a period of time. These results are presented in Figures 1 and 2. The oxygen level in the liquid very quickly dropped to zero. The amount of nitrate also decreased, eventually dropping to an undetectable level. The nitrate level remained below detectable limits even though a nitrified feed was continuously supplied to the culture. This indicated that a steady state condition was attained that matched the rate of nitrate reduction to the rate of NO_3^- replenishment.

The steady decrease in nitrate was accompanied by a slight increase of ammonia. The increase in ammonia concentration is difficult to explain, but other work has indicated that aerobic organisms placed in an anaerobic environment exhibit a slow leakage of amino acids into the surrounding medium [*Mechalas et al. 1970*]. These amino acids could serve as the ammonia source.

As a result of these experiments, a fourth statement can be added to the list of temporal events of the denitrification process:

4. Denitrification can proceed in the apparent absence of organic matter by utilizing the stored carbon reserves of microorganisms.

Hydrogen Sulfide

In a given water body below the photic zone, devoid of oxygen and nitrate, sulfate ion becomes the principal oxidant for respiration. This reaction results in the production of H_2S as an end product as illustrated in equation (4)

$$2CH_3CHOHCOOH + H_2SO_4 \rightarrow 2CH_3COOH + 2CO_2 + H_2S + 2H_2O \qquad (4)$$

Sulfate reduction is carried out by an unusual group of bacteria represented by the type species *Desulfovibrio desulfuricans*.

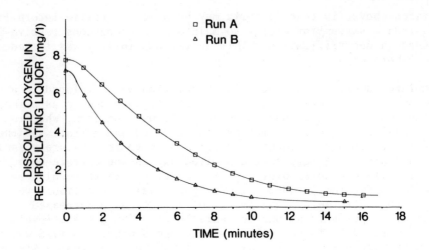

Fig. 1. Oxygen utilization in nitrifying system. System was operated aerobically just prior to run. At t = 0, the air and nutrient feed were turned off and the dissolved oxygen level was monitored.

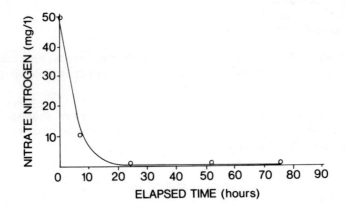

Fig. 2. Decrease in NO_3-N during denitrification. Continuous nutrient (NO_3-N) feed (residence time - 1000 minutes).

These organisms are obligate anaerobes requiring a reducing environment for active growth. They are generally regarded as heterotrophs requiring the presence of organic carbon for the synthesis of cell material [*Mechalas and Rittenberg, 1960; Postgate, 1960*].

A significant aspect of sulfate reduction, as contrasted to the denitrification process just described, is that the organic matter is not completely oxidized to CO_2 and H_2O. The starting substrate is only raised to the oxidation level of acetate. The sulfate reducing bacteria do not have the ability to metabolize beyond the acetate stage, and thus a great deal of potential energy is unavailable to these organisms.

In addition to the utilization of selected compounds as substrates for both cell synthesis and energy, *D. desulfuricans* can carry out a second set of reactions. It can derive energy from the oxidation of a nonassimilable organic molecule and use this energy to derive cell carbon from organic molecules which are unable to act as energy sources.

This was demonstrated in an experiment using isobutanol, a compound that can be quantitatively oxidized to isobutyric acid, but one which will not support cell growth [*Mechalas, 1959*]. Sulfate was required as the oxidant for this reaction. On the other hand, a complex substrate represented by yeast extract could support only minimal growth of the bacteria even in the presence of sulfate. When the two substrates were combined, however, normal growth occurred, with the production of isobutyric acid, hydrogen sulfide, and cell carbon.

The importance of this reaction in nature is that the spectrum of compounds that can be utilized by the sulfate reducing bacteria is significantly broadened. Thus, nonenergy yielding substrates can be assimilated if a suitable energy source is present.

D. desulfuricans is not solely dependent on organic materials for its energy metabolism. The sulfate reducing bacteria contain a hydrogenase enzyme system that enables them to couple the reduction of $SO_4^=$, using hydrogen gas, to the production of energy. This reaction is analogous to the role of isobutanol in the studies already mentioned above. It has been shown that H_2 can be used to replace the isobutanol in the yeast extract system, resulting in comparable levels of growth. Organic carbon is still required, however, for cell synthesis.

Thus, an overall reaction for sulfate reduction can be written as a two-step series: energy generation (5), and cell synthesis (6)

$$AH_2 + SO_4^= \rightarrow H_2S + A + energy \qquad (5)$$

$$(CH_2O) + energy \rightarrow cell\ material \qquad (6)$$

where AH_2 could either be the assimilated molecule, the oxidizable molecule, or molecular hydrogen.

As a result of the activities of sulfate reducing bacteria, sulfate ion is removed and hydrogen sulfide accumulates in the environment. However, free H_2S is seldom found in sediments as it is highly reactive and rapidly combines with metals contained in the clay minerals.

CO_2 is released simultaneously with the production of sulfide, and in the sulfate depleted waters reacts with calcium to form various carbonates. In addition, degradation of organic matter results in release of ammonia and phosphate, which may escape from the sediment and add nutrients to the overlying waters. Thus the activity of the sulfate reducing bacteria influences the physical and chemical properties of the sediment.

Methane

In an anaerobic environment devoid of sulfate, but containing organic matter, methane production becomes dominant. This gas has long been associated with the anaerobic digestion of organic compounds in fresh waters, and because this water is relatively sulfate free, the H_2S forming stage is either bypassed or is short in duration.

Although the mechanism of CH_4 production is not completely understood [*McCarty, 1964*], it has received a great deal of attention by virtue of its importance in the anaerobic digestion process as practiced in sewage treatment, and in the decomposition of plant materials and other organic wastes in flooded soils, marshes, cultivated paddy fields, and ocean sediments.

Anaerobic digestion is considered a two-stage process consisting of a nonmethanogenic phase (liquefaction) and a methanogenic phase (gas formation). In the first phase, a heterogeneous group of microorganisms converts proteins, polysaccharides, and lipids by the combined processes of liquefaction and fermentation into short chain alcohols and fatty acids, CO_2 and hydrogen gas. In the second stage, some of these organic acids and the CO_2 are converted to methane by a unique and metabolically limited group of anaerobic bacteria.

The complete anaerobic degradation of complex organic matter requires a diversity of enzymatic reactions, and involves a large number of bacterial types. Extracellular enzymes, apparently essential in the initial liquefaction stage, have been identified in anaerobic sewage digesters [*Agardy et al. 1963; Kotze et al. 1968; Thiel and Hattingh, 1967*].

An important point is that the second, or gas forming phase, requires simple organic molecules equivalent to the oxidation level of acetate as the starting substrate. In the deep sea environment, these compounds can result from the activities of the sulfate reducing bacteria which utilize the higher molecular weight organic acids and alcohols.

There has been some question as to whether certain species of methane forming bacteria could carry out the entire sequence from long chain organic acid fermentation to methane production. It is known that the long chain fatty acids first formed by lipid hydrolysis are oxidized through the mechanism of β-oxidation. When this process takes place under anaerobic conditions, an external hydrogen acceptor is required in order to reoxidize the electron transport coenzymes of the microorganism. Carbon dioxide (or bicarbonate), which has been produced in other metabolic processes, can serve as such an acceptor and would in turn be reduced to methane according to reaction (7) - (10)

$$CH_3(CH_2)_2COOH + 2H_2O \rightarrow 2CH_3COOH + 4H \tag{7}$$

$$2CH_3COOH \rightarrow 2CH_4 + 2CO_2 \tag{8}$$

$$1/2CO_2 + 4H \rightarrow 1/2CH_4 + 2H_2O \tag{9}$$

$$\overline{CH_3(CH_2)_2COOH \rightarrow 5/2CH_4 + 3/2CO_2} \tag{10}$$

However, there is now strong evidence that methanogenic bacteria are unable to utilize fatty acids above the level of acetate [*Toerien and Hattingh, 1969*]. Thus, the above CO_2 reduction mechanism for long chain fatty acid oxidation is probably not valid.

An alternative mechanism leading to the production of methane in anaerobic environments can be proposed. Many fatty acid oxidizing bacteria can directly oxidize their reduced electron transport system by producing H_2. This eliminates the need for an external electron acceptor, and the H_2 could then be used by the methane producing bacteria for a direct reduction of CO_2 according to equation (11) [*Barker, 1936*]

$$CO_2 + 4H_2 \rightarrow CH_4 + 2H_2O \tag{11}$$

Although the presence of large amounts of H_2 during the

anaerobic digestion of lipids by mixed populations of organisms
would be expected in the natural environment, this has not been demonstrated. Its generation in the nonmethanogenic phase, however,
is not excluded as many bacteria, including the sulfate reducers,
possess hydrogenase enzymes and the hydrogen would be rapidly utilized. If the H_2 producing phase does occur, the methane bacteria would also compete for this gas.

Recent work indicates that the actual methane forming bacteria
may be even more restricted in their nutritional requirements than
earlier studies indicate. The type species *Methanobacterium omelianskii* has been shown to be a symbiotic mixture of two bacterial types
[*Bryant et al. 1967*]. An intermediate metabolic type has been isolated that takes the simple organic compounds and fatty acids formed
in the hydrolytic stage and converts them to acetate. The acetate
formed in this reaction is the substrate for the methane producing
organism. The acetate-forming partner does not have the ability to
form methane.

The discovery that acetate may be the key intermediate in CH_4
formation in some environments has led to the re-examination of the
overall fermentation process. A succession of degradative steps
seems to be involved, each with a distinct microorganism population.
Complex organic substrates are successively broken down into smaller
and smaller fragments until the level of acetate is reached. This,
then, is the role of the nonmethanogenic phase.

Methanogenic Processes

The methanogenic, or gas forming, bacteria are obligate anaerobes that require a low redox potential for growth [*Bryant, 1965*].
There is evidence that the metabolic ability of these bacteria is
limited to acetate, methanol, formate, CO_2, and H_2 as substrates
for CH_4 generation.

Methane can result from two major metabolic pathways. One is
termed 'CO$_2$ reduction' and, as its name implies, involves the
direct hydrogenation of CO_2 to form CH_4. The second or 'acetic
acid fermentation' is an oxidation-reduction reaction, wherein part
of the acetate molecule is reduced to CH_4 and part to CO_2. This
reaction differs in some degree from the electron transfer reactions
previously discussed in that the same organic molecule acts as both
a hydrogen donor and hydrogen acceptor. Studies utilizing C^{14} and
deuterium labeled acetate, have shown that the methane in the acetate fermentation is formed directly from the methyl carbon and its
three attached hydrogens [*Koyama, 1963; Pine and Barker, 1956*].

The two pathways for methane formation are represented by equations (12) and (13)

$$CH_3COOH \rightarrow CH_4 + CO_2 \quad \text{(acetic acid fermentation)} \qquad (12)$$

$$CO_2 + 8H \rightarrow CH_4 + 2H_2O \quad \text{(CO}_2 \text{ reduction)} \qquad (13)$$

The CO_2 reduction theory was developed by van Niel, who postulated that all CH_4 was formed from the reaction of CO_2 with H_2. *Barker* [*1936*] expanded this theory to include the transfer of hydrogens from an organic molecule to reduce CO_2, as indicated in the following generalized equation

$$4H_2A + CO_2 \rightarrow 4A + CH_4 + 2H_2O \qquad (14)$$

Although it was originally conceived that simple alcohols could be utilized by the methanogenic bacteria, it is now thought that these organisms are restricted to the substrates, formate, acetate, methyl alcohol, carbon dioxide, and they require free hydrogen [*Bryant et al. 1967*]. This restriction postulates the presence of the intermediary population which produces hydrogen during the conversion of higher carbon number substrates.

The production of methane from ethanol would, therefore, require two distinct types of organism

$$2CH_3CH_2OH + 2H_2O \rightarrow 2CH_3COOH + 4H_2 \quad \text{(Bacterium 1)} \qquad (15)$$

$$CO_2 + 4H_2 \rightarrow CH_4 + 2H_2O \quad \text{(Bacterium 2)} \qquad (16)$$

The overall reactions involved in the complete methane fermentation of ethyl alcohol can be presented as the sum of the following reaction series

$$2CH_3CH_2OH + 2H_2O \rightarrow 2CH_3COOH + 8H \qquad (17)$$

$$8H + {}^*CO_2 \rightarrow {}^*CH_4 + 2H_2O \qquad (18)$$

$$2CH_3COOH \rightarrow 2CH_4 + 2CO_2 \qquad (19)$$

$$\rule{5cm}{0.4pt}$$

$$2CH_3CH_2OH \rightarrow {}^*CH_4 + 2CH_4 + CO_2 \qquad (20)$$

The $*CO_2$ required to drive this process would generally be available in all environments.

The reaction of equation (15) postulates the existence of a unique and heretofore unsuspected intermediate microbial population participating in the methane process. In a study of the relative importance of acetate fermentation versus CO_2 reduction [McCarty, 1964], it was determined that acetate fermentation could account for 60-100% of the methane formed, depending on the starting substrate. This implies that for acetate to accumulate, large amounts of molecular hydrogen may be produced during anaerobic digestion. As H_2 gas is noted in anaerobic digester gases in relatively small amounts, it must mean that the hydrogen is rapidly utilized and converted to methane by the microorganisms. The dynamics of this system are even more complex, for in the absence of CH_4 formation, H_2 still does not accumulate. This is tied into the observation [Bryant et al. 1967] that accumulation of extracellular hydrogen inhibits the growth of the hydrogen producing population. Thus, methane formation is required to remove the H_2 and permit the continued anaerobic degradation of small organic molecules.

In ocean sediments, the sulfate reducing bacteria may play a significant role by providing the acetate and CO_2 required by the methanogenic bacteria. They could also provide for a rapid uptake of excess H_2 by utilizing it to directly reduce the $SO_4^=$, and may, therefore, create an environment which will support a large H_2-producing population of bacteria prior to methane production.

Methane in the Natural Environment

Methane occurs in marine sediments under predictable conditions as discussed above. The environments are anaerobic (O_2 and NO_3^- are never present), highly reducing, and devoid of sulfates; they also contain significant amounts of organic matter, and are enriched in CO_2 or bicarbonate.

Organic rich marine sediments often provide environments that meet the above requirements, and relatively large amounts of methane can form there. Although the gas volumes in the sediments are difficult to quantify, Reeburgh [1969] reports CH_4 contents as high as 150 ml/1 in Chesapeake Bay. In ocean sediments, large amounts of methane may often be found below the zone of $SO_4^=$ reduction. The overlying bottom waters of anoxic basins, such as the Black Sea, the Cariaco Trench, Lake Nitinat, and other stranded fjords, contain rather large amounts of methane. In these waters a dramatic increase in dissolved gas is observed with increasing depth, especially away from the sill [Lamontagne et al. 1973]. Methane concentrations as high as 1.6 ml/1 have been reported for Lake Nitinat at a depth of 200 meters.

As one proceeds up the water column in these basins, the methane concentration rapidly decreases, with a minimum concentration being found at the depth where mixing occurs with normal aereated ocean waters. This is consistent with methane being generated in reducing sediments and taken up in the mixing zones. Near the surface, a second, but much smaller maximum, usually appears with an average concentration of 4.5×10^{-5} ml/l. The surface water is slightly supersaturated with respect to methane, based on calculations that the dissolved gas is in equilibrium with an atmospheric methane concentration of 1.38 ppm [*Lamontagne et al. 1973*].

The mechanism of formation of this small excess of methane in surface waters is not understood, but may be due to localized biological processes. Very little of the large quantities of methane generated and diffusing from the marine sediments is thought to ever reach the atmosphere. This is not too surprising when one considers the metabolic factors that control CH_4 accumulation and persistence. Methane can be oxidized to CO_2 and H_2O by a rather large number of bacteria. There is strong evidence that the CH_4 oxidation differs from the electron transport-dehydrogenation reactions previously discussed in that it is mediated through oxygenase enzymes that involve the direct incorporation of oxygen into the molecule. Thus, methane utilization would occur only in the aerobic layers of the ocean. In the deep water basins, or in the sediments, methane could survive for long periods of time. However, methane diffusing through the water column would be readily oxidized as it traversed the aerobic zone, and would seldom reach the air-water interface.

Anaerobic digestion on a global scale represents a very critical part of the carbon cycle. Vast amounts of vegetation decay or decompose in shallow waters represented by marshes, paddy fields, and submerged lands scattered over the earth's surface. The total methane production from terrestrial or fresh water environments has been estimated at 14.5×10^{14} g CH_4/yr [*Robinson and Robins, 1968*], with a lifetime of 1.5 years [*Weinstock and Niki, 1972*].

This gas may diffuse or bubble through shallow waters, with much of it escaping microbial oxidation before entering the atmosphere. It has been estimated that the atmosphere contains approximately 4.3×10^{14} g CH_4 [*Goldberg, 1951*]. All of these estimates agree that shallow water or soil environments are sufficient to maintain atmospheric levels, and that relatively little methane needs to escape from the deep oceans.

The processes that control the production of gases in marine sediments are predictable when based on a knowledge of microbial physiology. When organic matter is introduced into a water environment, a series of temporal changes occur, following each other in an ordered series of reactions. As substrates and environments change,

different microbial populations gain dominance, and in turn their
activities prepare the way for the succeeding population. Ultimately, the substrate is reduced to an oxidation state where, under
the conditions of the environment, useful metabolic energy can no
longer be derived and the final products may accumulate in the sediment.

REFERENCES

Agardy, F. J., R. D. Cole, and E. A. Pearson, Kinetic and activity
parameters of anaerobic fermentation systems, *SERL Rep. No. 63-2*,
University of California Berkeley, 1963.

Barker, H. A., On the biochemistry of the methane fermentation,
Arch. Mikrobiol., *7*, 1404, 1936.

Bryant, M. P., Rumen methanogenic bacteria, in *Physiology of
Digestion in the Ruminant*, edited by R. W. Dougherty, *et al.*,
pp. 411-418, Butterworths, Washington, 1965.

Bryant, M. P., E. A. Wolin, M. J. Wolin, R. S. Wolfe, and M. Mandel,
Methanobacillus omelianskii, a symbiotic association of two
bacterial species, *Bact. Proc.*, *19*, 1967.

Goldberg, L., The abundance and vertical distribution of methane
in the earth's atmosphere, *Astrophys. J.*, *113*, 567, 1951.

Goldhaber, M. B., and I. R. Kaplan, The sulfur cycle, in *The Sea*,
vol. 5, edited by E. D. Goldberg, 394 pp., Wiley Interscience,
New York, 1973.

Kotze, J. P., P. G. Thiel, D. T. Toerin, W. H. J. Hattingh, and
M. L. Siebert, A biological chemical study of several anaerobic
digesters, *Water Res.*, *2*, 195, 1968.

Koyama, T., Gaseous metabolism in lake sediments and paddy soils
and the production of atmospheric methane and hydrogen, *J. Geophys. Res.*, 3971, 1963.

Lamontagne, R. A., J. W. Swinnerton, V. J. Linnenbom, and W. D.
Smith, Methane concentrations in various marine environments,
J. Geophys. Res., *78*, 5317, 1973.

McCarty, P. L., The methane fermentation, in *Principles and
Applications in Aquatic Microbiology*, edited by H. Heukelekian
and N. C. Dondero, pp. 314-343, John Wiley, New York, 1964.

Mechalas, B. J., Energy coupling in Desulfovibrio desulfuricans,
Ph.D. thesis, University of Southern California, Los Angeles, 1959.

Mechalas, B. J., and S. C. Rittenberg, Energy coupling in Desulfovibrio desulfuricans, *J. Bact.*, *80*, 501, 1960.

Mechalas, B. J., R. H. Allen, III, and W. Matyskiela, A study of nitrification and denitrification, *EPA Rept. No. 17010 DR 07/70*, U. S. Government Printing Office, Washington, 1970.

Pine, M. J., and H. A. Barker, Studies on the methane fermentation - XII. The pathway of hydrogen in the acetate fermentation, *J. Bact.*, *71*, 644, 1956.

Postgate, J. R., On the autotrophy of Desulphovibrio desulphuricans, *Z. Allg.*, *Mikrobiol.*, *1*, 53, 1960.

Reeburgh, W. S., Observations of gases on Chesapeake Bay sediments, *Limnol. Oceanogr.*, *14*, 368, 1969.

Richards, F. A., and B. B. Benson, Nitrogen/argon and nitrogen isotope ratios in two anaerobic environments, the Cariaco Trench in the Caribbean Sea and Dramsfjord, Norway, *Deep-Sea Res.*, *7*, 254, 1961.

Robinson, E., and R. C. Robbins, Sources and abundances, and fate of gaseous atmospheric pollutants, *Stanford Res. Inst.*, *Proj. Rep. 6755*, Palo Alto, California, 1968.

Thiel, P. G., and W. H. J. Hattingh, Determination of hydrolytic enzyme activities in anaerobic digesting sludge, *Water Res.*, *1*, 191, 1967.

Toerien, D. T., and W. H. J. Hattingh, Anaerobic Digestion I. The microbiology of anaerobic digestion, *Water Res.*, *3*, 385, 1969.

Weinstock, B., and H. Niki, Carbon monoxide balance in nature, *Science*, *176*, 290, 1972.

DEPTH DISTRIBUTIONS OF GASES IN SHALLOW WATER SEDIMENTS

William S. Reeburgh and David T. Heggie

Institute of Marine Science
University of Alaska
Fairbanks, Alaska 99701

ABSTRACT

Depth distributions of Ar, N_2, CH_4, total CO_2 and total H_2S have been obtained in sediments from Chesapeake Bay, as well as fjords, coastal lagoons and lakes in Alaska. Interstitial water was separated from samples of sediment cores with a filter-press type sediment squeezer and transferred to a specially designed sampler-stripper, where the gases were stripped from solution and analyzed by gas chromatography. The technique permits repeated sampling of interstitial water from a given sediment sample, and yields unambiguous measurements of gas concentrations that are precise to about 5%.

In Chesapeake Bay, CH_4 increased from undetectable quantities at the sediment surface to concentrations of 150 and 85 ml/l in water depths of 30.4 and 15.2 meters. The observed maximum concentrations agree well with values calculated, assuming that the CH_4 concentration is controlled by ebullition from the sediment. The absence of CH_4 in the upper 25 cm of these sediments and the occurrence of Ar and N_2 in concentrations similar to the overlying water in this zone indicated mixing to at least this depth. Total CO_2 increased with depth to values of 1200 ml/l, pH was uniform with depth, and alkalinity increases to as high as 50 meq/l were observed. Low contents of total H_2S and an abundance of acid-labile sulfides in the sediments indicated that the concentration of H_2S is controlled by formation of solid phases.

In Alaska, methane was observed only in freshwater environments in concentrations far below those necessary for ebullition. Active denitrification was observed in Ace Lake sediments, where the N_2/Ar

ratio was 65 for one sample. Argon was constant with depth in all environments. Large quantities of H_2S were observed in the eel-grass beds at Izembek Lagoon. Total CO_2 distribution was similar in all of the fjords studied, showing an increase from about 50 ml/l at the sediment surface to relatively constant concentrations greater than 200 ml/l below about 30 cm.

INTRODUCTION

Large qualitative and quantitative differences are observed between the gases in surface waters and those present in sediments as interstitial waters. Because of these differences in composition, measurements of gases in sediments should be useful in interpreting the physical and chemical processes taking place beneath the sediment-water interface. For both environments, surface waters, and interstitial waters, the gas content is controlled by three factors: exchange across the interface, internal mixing, and biological processes. Because gases in sediments may be produced biologically within the sediments or mixed in from the overlying water, they offer opportunities to study both the chemical and physical processes in sediments. In interstitial waters, the presence of appreciable quantities of organic detritus and the large proportion of solid material with high specific surface area permits the growth of large and diverse bacterial populations that deplete the dissolved oxygen, causing anoxic conditions. Methane, CO_2, and H_2S are produced in the biochemical degradation of organic material and can be analyzed as gases. As sulfate reduction and methane production are performed by obligate anaerobes, depth distributions of these gases in sediments can be used to infer a limit of oxygen penetration. The inert gas Ar is present in measurable quantities and should be affected only by physical processes.

Several mathematical models of the sediment-water interface have been proposed by *Berner* [*1964*], *Duursma* [*1966*] and *Anikouchine* [*1966*]. Duursma's model considers diffusion only, and is oriented toward predicting the distribution of radionuclides introduced at the sediment surface. Berner and Anicouchine considered diffusion with no sorption, compaction, and chemical reaction in their models. The chemical reactions studied were ones with simple reaction kinetics for which experimental data were available. Berner considered sulfate reduction, and Anikouchine considered sulfate reduction, oxygen consumption, silica dissolution, and manganese dissolution as examples. Anikouchine's model also contained an advection term, which was shown to be small. Both Berner and Anikouchine showed that the diffusion and reaction terms were of approximately equal magnitude. All of the above models assume homogeneous undisturbed sediments, with uniform diffusion and the absence of mechanical mixing and bubbles.

Although these models represent a good start at understanding the processes controlling distributions of species in sediments, there are several processes that are known to have large effects that are not considered in the models. Biological mixing by burrowing and sediment ingesting organisms is an effective means of homogenizing the upper 10 cm or so of sediments. *Rhoads* [*1967*] and *Gordon* [*1966*] have measured sediment ingesting rates of polychaete worms, and *Haven and Morales-Alamo* [*1966*] have shown that oysters are able to mix the upper 6 cm of sediments. *Callame* [*1960*] has shown experimentally that dense water can convect quite rapidly into sediment filled with less dense water. Although Callame's experimental conditions were extreme, there appears to be confirmation of the process in the work of *Sanders et al.* [*1965*], who showed that sediments of the Pocasset River, a tidal estuary on Cape Cod, have higher salinities than the time-average salinity of the river water.

This paper describes measurements of depth distributions of Ar, N_2, CH_4, total CO_2 and total H_2S in a variety of shallow water environments, including Chesapeake Bay, and fjords, coastal lagoons, and lakes in Alaska. The work was undertaken to learn how varying degrees of physical mixing, biological activity, and sedimentation affect gas distributions in sediments. The work has been restricted to the upper two meters of sediments from these environments.

Several workers have captured and analyzed gas bubbles escaping sediments [*Shaw, 1914; Conger, 1943, Gould, 1960; Olsen and Wilder, 1961; Bean, 1969*]. These bubble studies are of qualitative interest only in that they yield no information about the depth distribution of gases in sediments, and represent samples altered in composition by passage through water overlying the sediments.

Previous quantitative studies of gases in sediments [*Emery and Hoggan, 1958; Koyama, 1953*] have involved extraction of gases from slurries of sediment samples and gas-free water by vacuum or stripping, followed by analysis by manometry, mass spectrometry or gas chromatography. Expression of the sediment gas results in terms comparable to the overlying water introduced difficulties in both of these investigations. Concentrations were obtained indirectly by either parallel determinations of sediment water content or by assuming the concentration of N_2 to be identical in the sediment and overlying water and using it as an internal standard. Duplication was not attempted, the data suggest contamination by atmospheric gases, and the distribution of samples was far from ideal.

In the present studies, interstitial water was separated from samples of sediment cores with a filter-press type sediment squeezer [*Reeburgh, 1967*] and transferred to a specially designed sampler-stripper unit, where the volume of interstitial water was measured. Gases were stripped from solution and analyzed, using gas chromatography in a manner similar to that described by *Swinnerton et al.*

[1962]. Gas concentrations in ml/l were determined directly. Contamination by air was minimized by loading the squeezers in a well-flushed glove bag filled with carrier gas and by making all subsequent transfers under pressure. The technique permits repeated sampling of interstitial water from a given sediment sample, and yields unambiguous measurements of gas concentrations that are precise to about 5% on natural samples. Details of the analysis are published elsewhere [Reeburgh, 1968].

The Chesapeake Bay work reported in this paper was accomplished by employing a Fisher model 29 gas partitioner that used helium as carrier gas. This instrument is equipped with a thermal conductivity detector. A Varian model 90-P3 gas chromatograph equipped with a Gow-Mac gas density balance detector was used in the Alaska work. Gow-Mac's Fil-The detector elements, consisting of a heated filament mounted upstream from a glass bead thermister and reported to be about twice as sensitive as thermisters or filaments used singly, were used in the gas density balance. With this detector, sample gases never contact the detector elements, and detector response is a predictable function of density or molecular weight, permitting use of a single gas for calibration. Using SF_6 as carrier, the sensitivity of this detector was observed to be 6-, 9- and 30-fold higher than the Fisher model 29 for Ar, N_2 and CH_4. However, analysis times were much longer, and the detector's increased sensitivity prevented measurements of Ar and N_2 on board ship. In the following figures, each point represents the mean of three determinations. The lengths of the vertical lines represent the depth interval of sediment sampled.

RESULTS

Two stations in Chesapeake Bay were occupied at different seasons. Station 858-C lies in the axial trough off the bay, and is occasionally exposed to anoxic bottom water during summer stratification. Station 858-D lies to the west of the axial trough, and is ordinarily exposed to oxic bottom water. Sediments at both stations are black muds. Water content determinations revealed no drastic discontinuities with depth at either station. Representative depth distributions of Ar, N_2, CH_4 and temperature at both stations under summer and winter conditions are shown in Figures 1 through 4. Figure 5 shows the depth distribution of total CO_2 and pH at station 858-D.

Ace Lake is located near the University of Alaska campus, and is believed to have formed as a result of thawing and subsidence of permafrost. Considerable background data are available for the lake as a result of studies by Barsdate [1968] and Clasby and Alexander [1970]. The lake is ice-covered from about October through June. Spring and fall circulation are confined to the upper 5 meters of

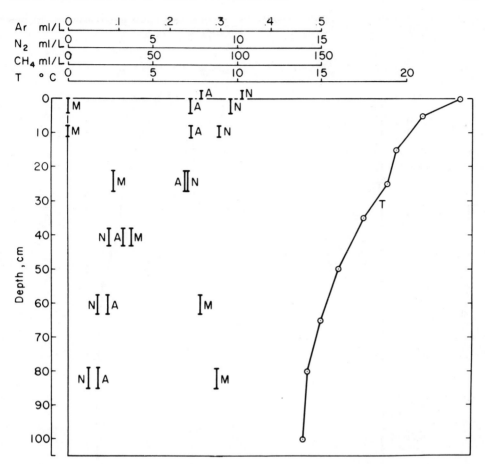

Fig. 1. Depth distributions of Ar, N_2, CH_4 (M) and temperature in Chesapeake Bay sediments, Station 858-D, 15.2 meters, 17-VIII-67.

the lake, and the deep waters remain anoxic continuously. Cores were taken from Ace Lake near the shore in water depths of about 1 meter. The upper parts of the cores contained a large amount of organic material. Below 16 cm, a marked texture change to a dry silty peat was observed. This material represents an example of the soil present before the development of the lake. Figure 6 and 7 show distributions of Ar, N_2 and CH_4, and total CO_2 in Ace Lake sediments.

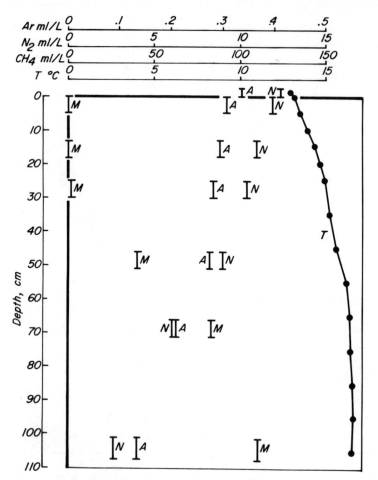

Fig. 2. Depth distributions of Ar, N_2, CH_4 (M) and temperature in Chesapeake Bay sediments, Station 858-D, 15.2 meters, 17-XI-66.

Gases were also studied in the sediments of several Alaskan fjords. Two Prince William Sound fjords, Blue fjord, a small outwash fan fjord with a deep sill, and Unakwik Inlet, a large outwash with a very pronounced sill, were studied. Queen Inlet, a large outwash fan inlet in Glacier Bay was also visited, but produced very sandy cores. In all of the fjord work, the ship's motion made operation of the gas density balance impossible at high sensitivity, and thus Ar and N_2 measurements were not obtained. Total CO_2 distributions in all fjords studied were similar; Figure 8 shows a typical depth distribution of CO_2 in a fjord environment.

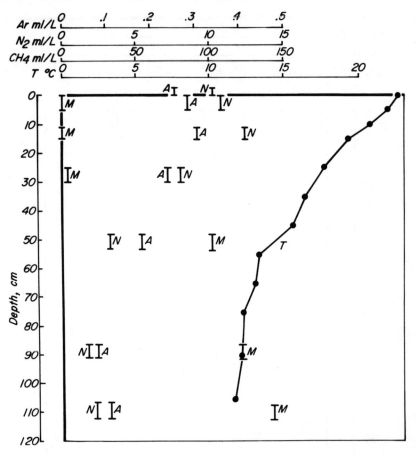

Fig. 3. Depth distributions of Ar, N_2, CH_4 (M) and temperature in Chesapeake Bay sediments, Station 858-C, 30.4 meters, 15-VIII-67.

Izembek Lagoon is one of many coastal embayments of the Bering Sea and is located near the western tip of the Alaska Peninsula. The shallow lagoon is separated from the Bering Sea by a series of barrier islands and exchanges water with the Bering Sea through tidal action. The lagoon supports large standing crops of eelgrass, *Zostera marina*. Large populations of migratory waterfowl feed here during August and September. The lagoon is ice-covered for variable periods during the winter. *McRoy* [*1970*] has studied the eelgrass and has described the environment. *McRoy and Barsdate* [*1970*] and *McRoy et al.* [*1972*] have studied the transport of phosphorus to and from the sediments by the eelgrass plants. This environment was

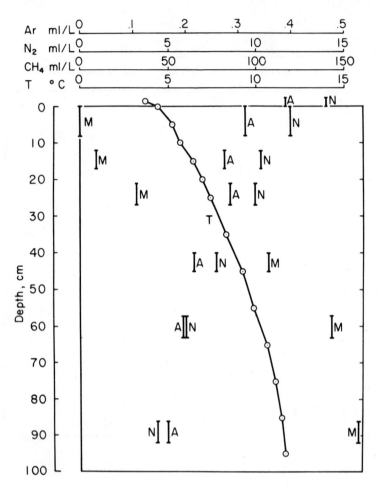

Fig. 4. Depth distributions of Ar, N_2, CH_4 (M) and temperature in Chesapeake Bay sediments, Station 858-C, 30.4 meters, 5-I-67.

studied to see what effects the rooted, vascular, eelgrass plants have on gases in sediments. The data are shown in Figures 9 and 10.

DISCUSSION

The processes governing the depth distribution of gases in Chesapeake Bay sediments have been discussed previously [*Reeburgh, 1969*]. Briefly, the Chesapeake Bay gas measurements show an absence

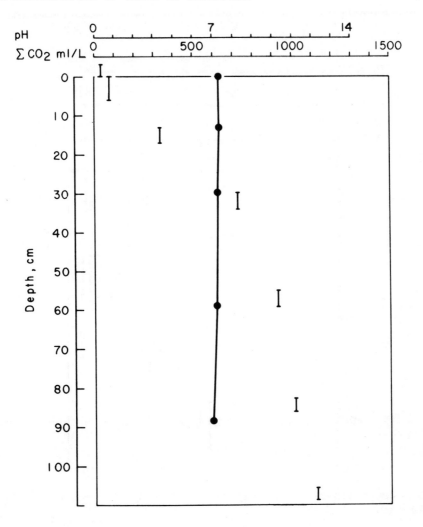

Fig. 5. Depth distributions of total CO_2 (bars) and pH in Chesapeake Bay sediments, Station 858-D, 15.2 meters, 15-V-68.

of CH_4 in the upper 20 cm of each core, followed by an increase to values that are dependent on water depth. Argon and N_2 are present in the region where CH_4 is absent in concentrations near that of the overlying water, and decrease to concentrations of about 0.1 and 2 ml/1 at a depth of 1 meter. The concentrations below 25 cm were well below the lowest summer concentrations of the overlying water. N_2/Ar ratios decrease with depth from values

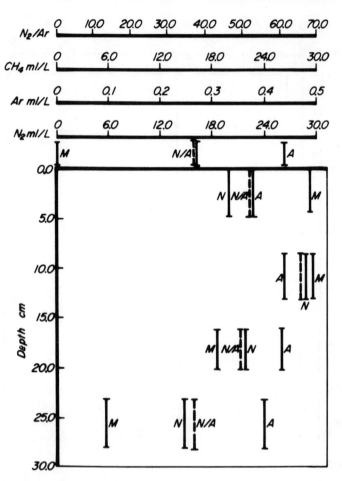

Fig. 6. Depth distributions of Ar, N_2, CH_4 and N_2/Ar in Ace Lake sediments, 1 meter, 3-VIII-71.

representing equilibration with the atmosphere [38:1, *Benson and Parker, 1961*] to values of 20:1 at 1 meter in the sediment, indicating selective removel of N_2. The removal of Ar can only be accounted for by a physical process.

Considering the question of whether CH_4 might be lost from sediments as bubbles, the ambient pressure at each station and CH_4 solubility coefficients were used to calculate critical concentrations as suggested by *Klots* [*1961*]. If the measured CH_4

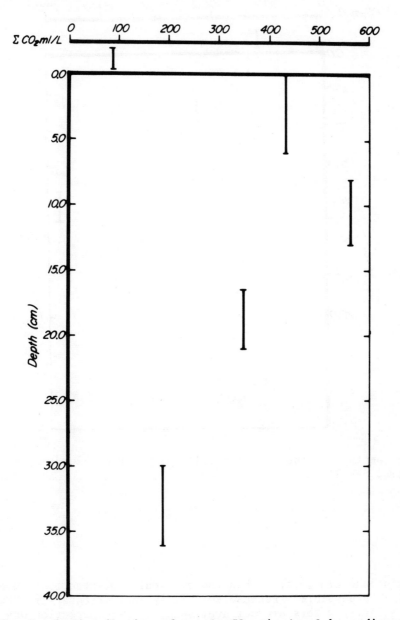

Fig. 7.　Depth distribution of total　CO_2　in Ace Lake sediments, 1 meter, 12-VIII-71.

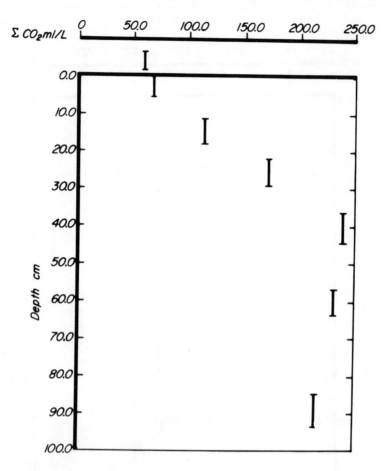

Fig. 8. Depth distribution of total CO_2 in Blue Fjord sediments, 80 meters, 6-V-71.

concentrations are greater than the critical concentrations, bubble formation is possible. Because solubility coefficients for CH_4 in low salinity waters are not available, a salt correction was applied to the distilled water solubility coefficients reported by *Winkler* [*1901*] (see *Atkinson and Richards*, [*1967*]). The solubility coefficients for Ar, N_2 and O_2 are similar in magnitude to CH_4 and decrease by about 1% per chlorinity unit; hence, this correction was applied to Winkler's values. The observed maximum CH_4 concentrations were within 10% of the calculated critical concentrations. Seasonal warming of 10°C would cause CH_4 in equilibrium

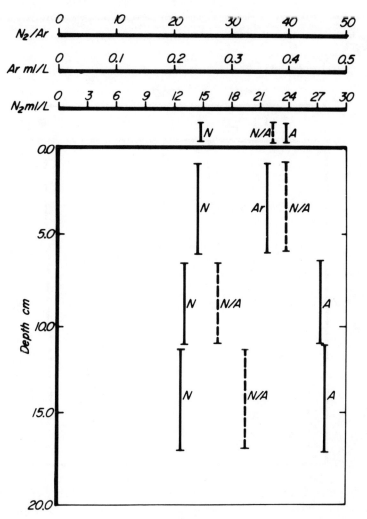

Fig. 9. Depth distributions of Ar, N_2 and N_2/Ar in Izembek Lagoon sediments, 0.8 meter, 11-VIII-71.

with water at a lower temperature to exceed the critical concentrations by about 30%, and thus it appears that bubbles could form and that ebullition from these sediments controls the maximum concentration of CH_4 in these sediments. No bubbles were observed along the walls of the coring tube or beneath the butterfly check valve of the corer.

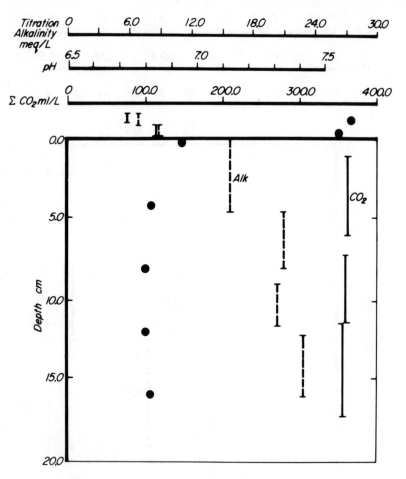

Fig. 10. Depth distributions of total CO_2, alkalinity and pH in Izembek Lagoon sediments, 0.8 meter, 11-VIII-71.

As both Ar and N_2 are depleted below 25 cm, stripping by the presumed CH_4 bubbles can be considered as a possible process leading to the observed distributions. The stripping process may be treated quantitatively by calculating the volume of gas phase required to contain the Ar and N_2 lost from the interstitial water and still be in equilibrium with the residual Ar and N_2 in the interstitial water. The volume of stripping gas obtained in this calculation is a minimum volume estimate, and implies an equilibrium stripping process if the calculated stripping volumes for both Ar and N_2 are equal. The Ar and N_2 data used in this calculation are from three cores taken at the same station and sampled at the surface and 1 meter depth [see Table I, *Reeburgh, 1968*]. The number of moles of Ar and N_2 lost to the gas phase for each liter of interstitial water was taken as the difference between the surface and 1 meter Ar and N_2 concentrations. The observed 1 meter Ar and N_2 concentrations were converted to the corresponding partial pressures using the appropriate solubility coefficients at 10°C and 10°/oo chlorinity. From the estimated number of moles lost and the partial pressures required for equilibrium, the stripping gas volumes for both Ar and N_2 were calculated. The stripping gas volumes for Ar and N_2 were, respectively, 80.3 and 81.3 ml/l of interstitial water. As uncertainties in the gas measurements are greater than the difference between the two estimates, the stripping volumes may be considered equal, indicating that Ar and N_2 are stripped from interstitial water in equilibrium ratios. The lower solubility of N_2 relative to Ar is the cause of its selective removal. Biological processes may affect the nitrogen distributions, but their effects appear to be overwhelmed by the CH_4 stripping process.

The occurrence of CH_4 at concentrations near the critical concentration and the indication of losses of CH_4 from the sediments, the failure to observe bubbles in the corer, and the indication of a process removing Ar and N_2 in equilibrium ratios has presented a puzzling situation until recently. The work of *Schubel* [*1974*] has shown that it is possible for bubbles of gas, presumably CH_4, to exist in fine-grained sediments near the present study area, where they affect the compressibility of the sediments and actively attenuate acoustical energy. The bubbles cannot be released by agitation, and only appear after the cores have stood on deck for some time. The existence and persistence of these bubbles provides some explanation of the equilibrium stripping process by allowing ample time for equilibration.

Some total CO_2 and pH measurements from Chesapeake Bay are shown in Figure 5. Carbon dioxide can be produced in sediments by both aerobic and anaerobic reactions. Dissolution of carbonate minerals in the sediments may account for a portion of the CO_2, but relatively large quantities of carbonate minerals must be dissolved to produce noticeable increases. Well-preserved mollusc shells were evident throughout these sediments. Considering the pH values and the observed alkalinities of 50 meq/l, it appears that the CO_2 must exist in solution predominantly as HCO_3^-. Unlike CH_4, which can escape the sediment as bubbles, the CO_2 remains in the sediments as an ionic species neutralized by cations produced or released in the sediments. In studies of interstitial and anoxic waters [*Berner et al. 1970; Gaines and Pilson, 1972*], ammonia has been shown to be partially responsible for the increased alkalinity observed in these environments and is probably responsible here.

Hydrogen sulfide was observed in concentrations less than 1 ml/l. The large quantities of acid-labile sulfide minerals present in the sediments indicate that the concentration of H_2S is controlled by precipitation.

Argon and N_2 in the upper portions of Chesapeake Bay sediment appear to track seasonal variations in the overlying water caused by temperature changes.

In the Alaskan work, Ace Lake proved to be the most dynamic environment studied. Methane was observed at concentrations up to about 30 ml/l at the sediment surface and decreased with depth. The maximum concentrations observed were below those necessary for ebullition. Perhaps the most striking feature of the Ace Lake gas distributions is the indication of active denitrification, shown by both the excesses of N_2 and the high N_2/Ar ratios. Critical concentrations for N_2 were exceeded in the 9-12 cm sample. Considering diffusion only and assuming a uniform diffusion coefficient of 0.5×10^{-5} cm^2/sec, a denitrification rate of 75 µg N/l/day is necessary at the 12 cm level to maintain the concentration gradient observed in Figure 6. This rate is comparable to that observed in the water column of Ace Lake by *Clasby* [*1972*] and in sediments of Lake Mendota [*Keeney et al. 1971*]. The total CO_2 distribution is shown in Figure 7 and can be seen to be similar in form to the N_2 distribution. All of the gas distributions in Ace Lake indicate that most of the biochemical activity is located above the 16 cm texture discontinuity. Below this level, the decreasing concentrations of CH_4 and N_2 indicate either decreased biological production or downward diffusion. The occurrence of CH_4 in the surface sample indicates that reaction terms are large compared to diffusion. Due to the shallow sediments, too little detail was obtained to permit conclusions regarding mixing by physical processes or burrowing organisms. Of the two processes, burrowing organisms, particularly *Chaoborus* or *Chironomus* larvae, would be expected to be most important due to the general absence of winds in interior Alaska. Hydrogen sulfide was not observed in these sediments, although small quantities are observed in the deep waters of Ace Lake. The low concentrations of sulfate and the high concentrations of iron in the lake waters probably produce this situation.

The Izembek Lagoon gas distributions are shown in Figure 9 and 10. The cores were taken within a stand of eelgrass in 0.8 meter of water and included plants along with their roots and rhizomes that extend down to 15 cm in the sediment. Hydrogen sulfide was observed in concentrations of about 1000 ml/l in the water just above the sediment surface. Measurements of H_2S were not possible in the sediments because of the high concentrations encountered. Methane was not observed in the sediments. Perhaps the most striking feature of these distributions is their uniformity with depth. Argon and N_2 in the sediments are essentially identical with the overlying water. The lack of excess N_2 or increased N_2/Ar ratios indicate that denitrification has not occurred. Total CO_2 was measured in the fluid squeezed from eelgrass leaves, and was found to be similar to the concentrations encountered in the sediments. These dense stands of eelgrass restrict water circulation and serve as sediment traps. Sediments beneath the stands of eelgrass consist largely of silt- and clay-size material combined with a large fraction of decaying eelgrass fragments. Although the uniform distributions might be caused by the restricted circulation of water within the eelgrass stand, they might be the result of transport of gases to and from the sediments by the vascular plants. *Teal and Kanwisher* [*1966*] have studied transport of O_2 and CO_2 in *Spartina* and found that calculated fluxes of these gases determined from concentration gradients agreed well with measured fluxes. Surrounding the leaves of the plants with a solution containing either a tracer or a gas foreign to the system like helium or SF_6, followed by measurements on the sediment, might be useful in determining whether the uniform distributions observed are caused by the plants.

McKinney and Conway [*1957*] predicted that the products of denitrification, sulfate reduction and methane production should appear sequentially as each of the specific substrates is depleted. Their predictions are followed in these environments. Where $SO_4^=$ is low or a mechanism for removing H_2S is available, as in Chesapeake Bay and Ace Lake, CH_4 was observed. No CH_4 was observed in Izembek Lagoon, where sulfate reduction was taking place.

ACKNOWLEDGMENTS

This work was supported by U.S. Atomic Energy Commission Contract AT(30-1)-3497 to the Johns Hopkins University, and National Science Foundation Grant GA 19380 to the University of Alaska.

Contribution No. 210, Institute of Marine Science, University of Alaska.

REFERENCES

Anikouchine, W. A., A model of the distribution of dissolved species in interstitial water in clayey sediments, Ph.D. thesis, Univ. of Washington, 1966.

Atkinson, L. P., and F. A. Richards, The occurrence and distribution of methane in the marine environment, *Deep-Sea Res.*, *14*, 673, 1967.

Barsdate, R. J., Pathways of trace elements in arctic lake ecosystems, *Progress Report to U. S. Atomic Energy Comm.*, *Contract AT(04-3)3I9PA4*, Inst. of Marine Science, Univ. of Alaska, 1968.

Bean, D., Seasonal variation of free gas ebullition from lake sediments, Ph.D. thesis, Univ. of Rhode Island, 1969.

Benson, B. B., and P. D. M. Parker, Nitrogen/argon and nitrogen isotope ratios in aerobic sea water, *Deep-Sea Res.*, *7*, 237, 1961.

Berner, R. A., Distribution and diagenesis of sulfur in some sediments from the Gulf of California, *J. Mar. Geol.*, *1*, 117, 1964.

Berner, R. A., M. R. Scott, and C. Thomlinson, Carbonate alkalinity in the pore waters of anoxic marine sediments, *Limnol. Oceanogr.*, *15*, 544, 1970.

Callame, B., Etude sur la diffusion des sels entre les eaux surnageantes et les eaux d'imbibition dans les sediments marins littoraux, *Bull. Inst. Oceanogr.*, *1181*, 1960.

Clasby, R. C., Denitrification in a sub-arctic lake, M.S. thesis, Univ. of Alaska, 1972.

Clasby, R. C., and V. Alexander, Rates of denitrification in the anoxic zone of a subarctic lake, in *Dynamics of the Nitrogen Cycle in Lakes*, pp. 53-64, U. S. Natl. Tech. Inform. Serv., P. B. Rept. 203791, 1970.

Conger, P. S., Ebullition of gases from marsh and lake water, *Chesapeake Biol. Lab. Publ.*, *59*, 1943.

Duursma, E. K., Molecular diffusion of radioisotopes in interstitial waters of sediments, in *Disposal of Radioactive Wastes into Seas, Oceans, and Surface Waters*, pp. 355-371, International Atomic Energy Agency, Vienna, 1966.

Emery, K. O., and D. Hoggan, Gases in marine sediments, *Amer. Ass. Petrol. Geol. Bull.*, *42*, 2174, 1958.

Gaines, A. G., Jr., and M. E. O. Pilson, Anoxic water in the Pettaquamscutt River, *Limnol. Oceanogr.*, *17*, 42, 1972.

Gordon, D. C., Jr., The effects of the deposit feeding polychaete, *Pectinavia gouldii* on the intertidal sediments of Barnstable Harbor, *Limnol. Oceanogr.*, *11*, 327, 1966.

Gould, H. R., Character of the accumulated sediment, gas content, in *Comprehensive Survey of Sedimentation in Lake Mead, 1948-1949*, pp. 180-181, U. S. Geol. Survey Prof. Paper 295, 1960.

Haven, D. S., and R. Morales-Alamo, Use of fluorescent particles to trace oyster biodeposits in marine sediments, *J. Cons., Cons. Perma. Int. Explor. Mer.*, *30*, 267, 1966.

Keeney, D. R., R. L. Chen, and D. A. Graetz, Importance of denitrification and nitrate reduction in sediments to the nitrogen budgets of lakes, *Nature*, *233*, 66, 1971.

Klots, C. E., Effect of hydrostatic pressure upon the solubility of gases, *Limnol. Oceanogr.*, *6*, 365, 1961.

Koyama, T., Measurement and analysis of gases in sediments, *J. Earth Sci., Nagoya Univ.*, *1*, 107, 1953.

McKinney, R. E., and R. A. Conway, Chemical oxygen in biological waste treatment, *Sewage Ind. Wastes*, *29*, 1097, 1957.

McRoy, C. P., On the biology of eelgrass (*Zostera marina* L.) in Alaska, Ph.D. thesis, Univ. of Alaska, 1970.

McRoy, C. P., and R. J. Barsdate, Phosphate absorption in eelgrass, *Limnol. Oceanogr.*, *15*, 6, 1970.

McRoy, C. P., R. J. Barsdate, and M. Nebert, Phosphorus cycling in an eelgrass (*Zostera marina* L.) ecosystem, *Limnol. Oceanogr.*, *17*, 58, 1972.

Olsen, F. C. W., and B. Wilder, Gases in bottom sediments, *Bull. Mar. Sci. Gulf. Carib.*, *11*, 207, 1961.

Reeburgh, W. S., An improved interstitial water sampler, *Limnol. Oceanogr.*, *12*, 163, 1967.

Reeburgh, W. S., Determination of gases in sediments, *Environ. Sci. Technol.*, *2*, 140, 1968.

Reeburgh, W. S., Observations of gases in Chesapeake Bay sediments, *Limnol. Oceanogr.*, *14*, 368, 1969.

Rhoads, D. C., Biogenic reworking of interstitial and subtidal sediments in Barnstable Harbor and Buzzards Bay, Massachusetts, *J. Geol.*, *75*, 461, 1967.

Sanders, H. L., P. C. Mangelsdorf, Jr., and G. R. Hampson, Salinity and faunal distribution in the Pocasset River, Massachusetts, *Limnol. Oceanogr.*, *10*, (Suppl.), R216, 1965.

Schubel, J. R., Gas bubbles and the acoustically impenetrable, or turbid, character of some estuarine sediments, in *Natural Gases in Marine Sediments*, edited by I. R, Kaplan, pp. 275-298, Plenum Press, New York, 1974.

Shaw, E. W., Gas from mud lumps at the mouth of the Mississippi, *U. S. Geol. Survey Bull.*, *591A*, 21, 1914.

Swinnerton, J. W., V. J. Linnenbom, and C. H. Cheek, Determination of dissolved gases in aqueous solutions by gas chromatography, *Anal. Chem.*, *34*, 483, 1962.

Teal, J. M., and J. W. Kanwisher, Gas transport in the marsh grass, *Spartina alterniflora*, *J. Exp. Bot.*, *17*, 355, 1966.

Winkler, L. W., Die Löslichkeit der Gase in Wasser, *Ber. Deut. Chem. Gasell.*, *34*, 1408, 1901.

METHANE AND CARBON DIOXIDE IN COASTAL MARSH SEDIMENTS

Thomas Whelan, III

Coastal Studies Institute
Louisiana State University
Baton Rouge, Louisiana 70803

ABSTRACT

In order to better understand processes of diagenetic gas production in organic-rich sediments, CH_4 and CO_2 were analyzed, using two extraction techniques, in several environments of the South Louisiana marshlands. Investigation of twelve surface cores and two 100-cm cores indicated that total methane ranged in concentration from 0.12 to 12.2 µl/gm dry sediment, and dissolved methane ranged from 0.13 to 7.4 µl/ml interstitial water. Total CO_2 in the interstitial water ranged from 224 to 410 µl/ml IW. Carbon isotope ratios were determined on CO_2 released from sediment under vacuum. Dissolved organic carbon in the interstitial water of surface sediments ranged from 15.0 to 38.0 mg C/l. The environmental variation and mechanism of methane production are discussed.

INTRODUCTION

Investigations concerning the origin and distribution of sedimentary gas have become important recently for several reasons. Shallow CO_2 and CH_2 production, initiated by degradation of sedimentary organic matter, is known to be an important link in certain diagenetic processes such as carbonate cementation [*Garrison et al. 1969; Allen et al. 1969; Whelan and Roberts, 1973; Nissenbaum et al. 1972*].

It has been suggested that the concentration of CH_4 in estuarine interstitial waters is regulated by bubble formation followed by removal from the sediment column [*Reeburgh, 1969*]. The existence of gas bubbles in sediment has been verified by anomalous acoustical behavior caused by the difference in density between gas and

47

sediment [*Jones et al. 1964; Levin, 1962*]. *In situ* gas generation may have a pronounced effect on sediment stability, directly by entrapment of gas bubbles in unconsolidated sediment or indirectly by initiation of certain diagenetic features such as carbonate cementation and nodule formation. Both phenomena may result in erroneous interpretation of the acoustical properties in marine sediments.

Several theories have been proposed for the production of shallow methane gas, including direct fermentation of organic matter by microorganisms [*Koyama, 1964*] and reduction of preformed CO_2 by molecular or organically available hydrogen [*Nissenbaum et al. 1972*]. Generally, high production of CH_4 occurs in sedimentary environments where organic matter is abundant. Marsh environments are typified by shallow-water, organic-rich sediments with high gas content. These sediments provide an ideal framework for this preliminary investigation concerning the origin and distribution of sedimentary marsh gas. The purpose of this paper is to investigate the distribution and relative quantities of CH_4 and CO_2 gases in the coastal marsh of southern Louisiana, and to gain insight into the mechanism of CH_4 formation.

STUDY SITE AND METHODS

Location

Samples were taken from the marsh environments associated with the Barataria Bay complex of southern Louisiana (Figure 1) during November 1972 and September 1973. The southern sector of the marsh is brackish; it is flushed with sea water during tidal changes. The northern sector has fresher waters and especially organic-rich sediment; frequently peat is found in the surface layers. *Spartina alterniflora* and *S. patens*, the most common plants, are the most important source of organic matter to the sediment. Montmorillonite is the predominant clay mineral [*Ho, 1971*].

All core samples, except AL-1P and AL-2P, were taken in 1-1.5 meters of water in interdistributary lakes surrounded by marsh. These sediments remain under water throughout the year. Samples designated as Al-1P and Al-2P were taken directly through the marsh where less than 20 cm of water was present.

Coring Technique

Piston cores of approximately 60-70 cm and 100-115 cm were taken in 7.6-cm PVC core barrels. Each core was quickly sealed with a rubber stopper before it was removed from the water in order to minimize gas loss and air contamination. After pistons and rubber

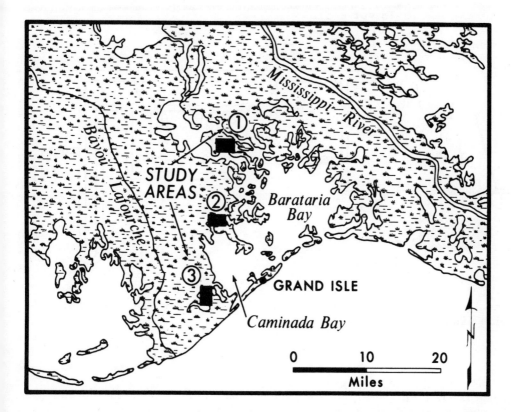

Fig. 1. Location map and sampling area in the Barataria Bay marsh system.

stoppers for each core were secured in place, the cores were stored on ice in the field and kept at 4°C until laboratory analysis (less than three weeks).

Gas Analysis

Sedimentary methane and carbon dioxide were analyzed by two separate methods. Gas from the short cores taken in November was analyzed by the system shown in Figure 2. The piston is held in place by the cable and pully system while the 2.6-liter chamber is evacuated to less than 1 cm Hg. The vacuum is closed and the core extruded into the vessel by releasing tension on the piston. Purified He gas is used as the carrier to strip gas from the chamber at a flow rate of 0.5-1 liter/min for 15 minutes. The gas is first routed through a 15-cm activated charcoal trap at -70°C, which

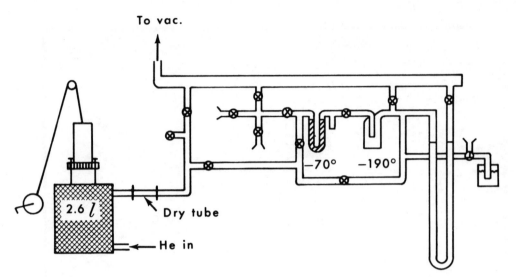

Fig. 2. Gas extraction system for isolating CH_4 and CO_2. Methane is absorbed in the charcoal trap at -70°C and CO_2 is condensed with liquid N_2 at -190°C.

absorbs CH_4 [*Swinnerton and Linnenbaum, 1967*], and then routed through a liquid nitrogen trap at -190°C to condense CO_2.

Gas samples were removed from the manifold through a gas-tight septum and analyzed by gas chromatography isothermally at 35°C on a 6-ft x 1/8-inch stainless-steel column packed with Chromosorb 102. A thermal conductivity detector was used, and the integrated peak areas were multiplied by appropriate response factors and corrected for air contamination. Replicate analysis of standard gas mixtures isolated on the extraction apparatus showed ±10% and ±15% reproducibility for CO_2 and CH_4, respectively. The gas content of sediments as determined by this method includes gas present as bubbles as well as dissolved gas.

Dissolved CH_4 and CO_2 were determined in the interstitial water of two longer cores and three shorter cores from sites AL in the southern sector, and JTF in the northern sector of the marsh. Intervals of approximately 10 cm were extruded from each core and the pore water squeezed into airtight 10-ml plastic syringes. Ten ml of pure water was injected into a helium-purged gas-stripping device similar to that described in *Reeburgh* [*1968*]. Dissolved methane was stripped from the water by purging with He for 5 minutes. The gases were routed through an activated charcoal trap cooled with dry ice-acetone to quantitatively absorb methane. The trap was heated to 90°C with a hot-water bath to release CH_4, and a 1-cm^3 sample was injected into the gas chromatograph. Replicate analysis of standard methane mixtures isolated by this method showed ±3% reproducibility.

Total CO_2 was determined on separate samples of interstitial water by infrared adsorption of CO_2 released from acidified samples of interstitial water (Oceanography International Total Carbon System).

Dissolved organic carbon was determined for the same sediment from which gas was extracted; the method used, described by *Fredricks and Sackett* [1970], was wet oxidation at 175°C for 24 hours with potassium persulfate in sealed glass ampoules.

The interstitial water was pressed from the sediment according to the method of *Reeburgh* [1967].

Carbon isotope determinations were made on the CO_2 gas extracted from the sediment using the first method. Results are reported as $\delta^{13}C$ values, defined by the expression

$$\delta^{13}C \ (^{\circ}/_{\circ\circ}) = \left[\frac{\left[^{13}C/^{12}C \right]_{sample} - \left[^{13}C/^{12}C \right]_{std}}{\left[^{13}C/^{12}C \right]_{std}} \right] \times 10^3$$

The reference standard was the PDB limestone.

RESULTS

Nine short cores were taken in November 1972 on a north-south line through the marshland of the Barataria Bay system. Two longer cores (100-115 cm) and three 60-70 cm cores were taken in September 1973 in order to investigate the depth distribution of methane and carbon dioxide, and to confirm previous results using an alternate method of gas analysis.

Study areas 3, 2, and 1 (Figure 1) identify the general sampling areas AL, BJ, and JTF, respectively. The samples taken within each area were approximately 50 meters apart. The single sample LL was taken from an open brackish water bay several miles north of JTF. The sediments of this area frequently contain shell material from the mollusc *Rangia*.

The chlorinity of the column water and interstitial water decreased northward, as shown in Table 1. The ratio CO_2/CH_4 also decreased northward. Dissolved organic carbon ranged from 15 to 38 mg C/l in the interstitial water, and from 10.2 to 28.4 mg C/l in the water column above the sediment. The concentrations of CO_2 and CH_4, as indicated in Table 1, include gas present as bubbles as well as a partial contribution of gas released from solution due to degassing of the pore water under vacuum.

Table 2 demonstrates the variability of dissolved CH_4 from separate cores taken in each of the three major study areas. Core AL-10 contains three to four times less dissolved CH_4 than AL-11. Dissolved CH_4 in JTF-4 and JTF-6 are very similar; these cores contain approximately 20-100 times more CH_4 than cores from the other two areas. Total CH_4, as determined by the vacuum degas method, is higher than dissolved CH_4 in each case. The values reported in the last column of Table 2 were averages of the concentrations of CH_4 reported in Table 1 for AL-3 and AL-4, BJ-1 and BJ-2, and JTF-1 and JTF-2.

The relationship between dissolved organic carbon in the interstitial water and the concentrations of CH_4 and CO_2 from the short cores, which were analyzed by the vacuum degas method, is shown in Figure 3. The straight line demonstrates the trend of decreasing CO_2 with increasing interstitial dissolved organic carbon.

Figures 4 and 5 illustrate the profile of ΣCO_2 and dissolved CH_4 concentrations as a function of depth within the sediment in the AL and JTF areas, respectively. Methane concentrations increased with depth in core AL-11, from 0.2 µl/ml IW near the sediment surface to 1.4 µl/ml IW at 112 cm. Total CO_2 increased from 364 µl/ml IW at the sediment surface to 410 µl/ml IW at 112 cm.

Dissolved methane concentrations in core JTF-4 were an order of magnitude higher at the surface than found in the AL-11 core (3.1-7.4 µl/ml IW). A slight increase in methane to 65 cm was observed. Both total CO_2 and CH_4 dropped below surface sediment concentrations at 75-85 cm.

DISCUSSION

Decomposition of sedimentary organic mater has been proposed by various investigators as a necessary requirement for gas generation in Recent freshwater and marine sediments. Marsh environments are typified by organic-rich sediments and high gas content. *Emery and Hoggan* [1958] reported CH_4 concentrations of 0.24 and 0.047 ml CH_4/l interstitial water from two separate marsh sediments. In Chesapeake Bay, CH_4 dissolved in interstitial water increased with sediment depth to a maximum value of 150 ml/l [*Reeburgh, 1969*].

TABLE 1. Some Geochemical Parameters of Marsh Column (CW) Water and Interstitial Water (IW)

	Chlorinity (ppt)		Total Organic Carbon (mg C/l)		CO_2 (μl/gm dry sed.)	CH_4	CO_2/CH_4	$\delta^{13}C$ $^o/_{oo}$ CO_2 (PDB)
	IW	CW	IW	CW				
AL-1P	12.6		25.4		4.1	0.9	4.5	-19.3
AL-2P	13.4	13.3	21.6	10.2	12.2	1.2	10.1	
AL-3	13.5		26.0		7.3	0.12	60.8	-18.0
AL-4	13.3	13.8	15.0	11.6	43.8	0.46	95.2	-19.8
BJ-1	12.0		22.0		0.41	0.16	2.56	
BJ-2	11.7	11.9	24.0	17.2	6.45	0.44	14.7	
JTF-1	6.5		31.6		1.83	4.25	0.43	-23.2
JTF-2	6.4	6.5	38.0	28.4	n.d.	12.30	-	-21.7
LL-1	7.3	7.5	25.0	19.8	6.15	4.50	1.37	-12.9

TABLE 2. Comparison of Total CH_4 (Vacuum Degas Method) and Dissolved CH_4 in Marsh Sediments

Sample Designation	Dissolved CH_4 $\mu l/gm$ sed.*		Total CH_4 $\mu l/gm$ sed.
	7.6 cm	30.5 cm	0-60 cm
AL-10	0.03	0.05	-
AL-11	0.13	0.22	-
AL-3 and AL-4[†]	-	-	0.29
BJ-7	0.19	0.16	-
BJ-1 and BJ-2[†]	-	-	0.36
JTF-4	3.1	5.2	-
JTF-6	3.7	3.4	-
JTF-1 and JTF-2[†]	-	-	8.3

*Assuming these sediments contain 50% water by weight, the values reported for methane in μl CH_4/ml IW can be expressed as μl CH_4/gm dry sediment.

[†]Average of CH_4 concentration from cores taken in November 1972; all other data reported here were taken in September 1973.

It was proposed that the concentration of CH_4 is regulated by ebullition from the sediment as bubbles. Comparison of the total and dissolved methane concentrations in Table 2 indicates that a significant fraction of the total methane concentration in the marsh exists as gas bubbles trapped in the sediment column. Seasonal differences in the sediment temperature between November 1972 and September 1973 were too small to account for the difference in dissolved and total methane reported in Table 2.

Dissolved CH_4 increased from 0.2 at the surface to 1.4 $\mu l/ml$ at 110 cm for area AL in the southern marsh, as shown in Figure 4. Methane at this depth within the sediment was two orders of magnitude less than observed in Chesapeake Bay sediments at the same depth, even though the sedimentary organic carbon is higher in the marsh (10% marsh versus 3% Chesapeake Bay). The hydrostatic head, however, was much less when considering that the samples reported

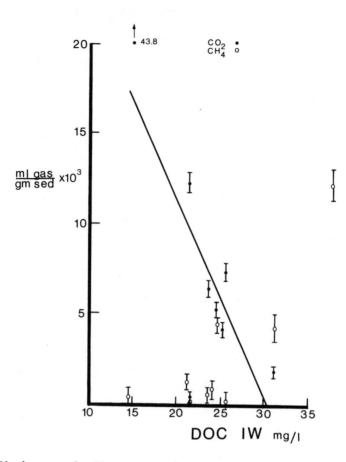

Fig. 3. Methane and CO_2 concentration in marsh sediments as a function of dissolved organic carbon in the interstitial water.

here were taken in only 1-1.5 meters of water as opposed to 15 and 30 meters in the Chesapeake Bay study. A similar trend with depth was observed in the dissolved CH_4 concentration in the northern JTF area (Figure 5), but CH_4 concentrations were approximately ten times higher than in the southern area. A sharp decrease in total CO_2 and CH_4 occurred from 65 to 85 cm in this core. This may be explained by a change in environment of deposition. Laminar bedding of alternate layers of clay and silt were noted at this interval in the core. More geological and chemical information will be required in order to explain this anomalous trend in methane and total CO_2 distribution.

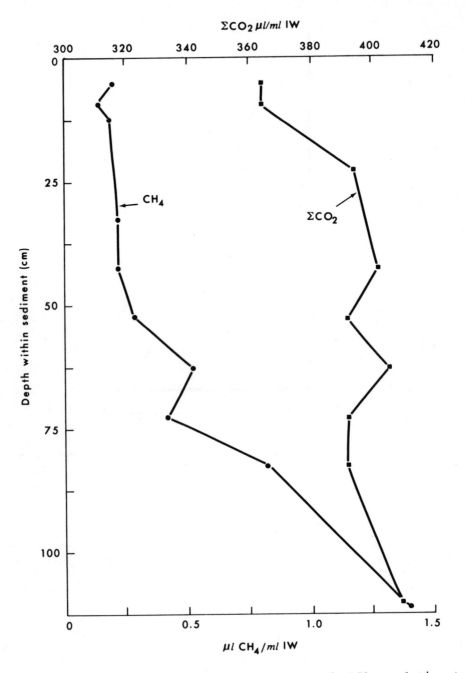

Fig. 4. Concentration of dissolved CH_4 and ΣCO_2 relative to depth for station AL-11 (southern area), 8 September 1973.

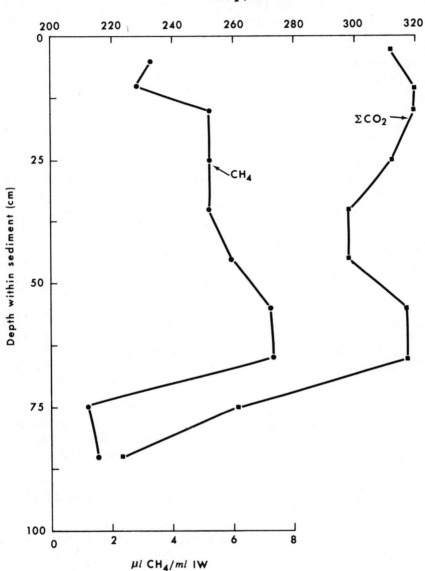

Fig. 5. Concentration of dissolved CH_4 and ΣCO_2 relative to depth for station JTF-4 (northern area), 8 September 1973.

It is interesting to note that the ΣCO_2 concentration approximately correlates with the trend of dissolved CH_4 in cores JTF-4 and AL-11. Total CO_2 is about two and a half times lower in the southern marsh at 100 cm than reported for estuarine sediments [$Reeburgh,\ 1969$] and two to five times higher than reported for nearshore marine sediments [$Presley\ and\ Kaplan,\ 1968$]. It is apparent that the contribution to the ΣCO_2 from organic matter decomposition is more uniform with depth in marsh sediments than in other marine environments and that the processes which regulate methane production are associated with those which yield biogenic CO_2. More information concerning the chemistry of interstitial water (such as the concentrations of SO_4^{-2}, S^{-2}, and dissolved organic carbon) is needed in order to account for the precise mechanism(s) of methane generation in this environment.

Gas concentration, determined by the vacuum degas technique, as a function of dissolved organic carbon (DOC) of the interstitial water is shown in Figure 3. Carbon dioxide gas (including some dissolved HCO_3^- that was liberated as CO_2 under vacuum during extrusion of the core) decreased with increasing interstitial DOC, possibly as a result of the reduced amount of organic matter decomposition by sulfate-reducing bacteria in the northern marsh, as reported by $Hood\ and\ Colmer$ [1971]. It has been suggested, on the basis of several lines of evidence, including the carbon isotope composition of the total CO_2 in interstitial waters, that CH_4 can be generated from the reduction of CO_2 according to the reaction [$Nissenbaum\ et\ al.\ 1972;\ Koyama,\ 1964$]

$$CO_2 + 4H_2 \rightarrow CH_4 + 2H_2O$$

Laboratory experiments on bacterial degradation of organic matter in soil have demonstrated that the production of CO_2 gas is accompanied by release of H_2S from sulfate reduction; this is followed, after a short time, by the appearance of CH_4 and hydrogen [$Takai\ and\ Kamura,\ 1966$]. A competition may exist in marine sediments for hydrogen between sulfate-reducing bacteria and methane-producing bacteria. If CH_4 production begins as the sulfate reduction diminishes, it could be expected that the areas of low chlorinity (sulfate) and high dissolved organic matter in the sediment pore water would have the capability of producing CH_4 by this mechanism.

The carbon isotope composition of CO_2 gas extracted by the vacuum technique is more positive in the southern area, where CO_2 concentration is higher, than in the northern area, where CO_2 is lower and CH_4 is higher. This appears to be consistent with the trend reported for total dissolved CO_2 in the interstitial waters of a reducing fjord [$Nissenbaum\ et\ al.\ 1972$], however, in the present study the range of $\delta^{13}C$ values is much more narrow. *Nissenbaum et al.* proposed that the carbon isotope composition of the ΣCO_2 in interstitial waters was determined by the kinetic isotope effect associated with the reduction of CO_2 to CH_4. The difference in $\delta^{13}C$ values reported here could be due to a greater contribution of terrestrial organic matter to the sediment in the northern area as opposed to the southern area. The $\delta^{13}C$ value -12.9% for station LL-1 (Table 1) was anomalously positive. In this sample a large amount of shell material (*Rangia*) was noted. Solution of shell carbonate would shift the carbon isotope ratio of sedimentary CO_2 in a positive direction with respect to organic matter, and hence the $\delta^{13}C$ value in this sample might not necessarily be determined by the fractionation associated with reduction of CO_2 to CH_4.

SUMMARY

The Louisiana marsh provides a unique environment in which to study the distribution and origin of sedimentary gas. Methane concentrations were an order of magnitude higher in the sediments of the northern area, which is associated with organic-rich and low-chlorinity pore water, than in the southern area, which is influenced by tidal flushing and more saline water. By using two techniques of gas extraction, evidence was obtained that a significant fraction of the methane exists as gas bubbles trapped within the sediment. Total CO_2 and dissolved methane demonstrated a similar variation with depth in each of two cores taken from the northern and southern areas. The relationship between total CO_2, dissolved methane, and the carbon isotope composition of sedimentary CO_2 suggests that the presence of CH_4 may be determined by chemical reduction processes in the pore waters, although the maximum concentration may be regulated by bubble formation.

ACKNOWLEDGMENTS

I wish to thank Dr. Brian Eadie of Texas A & M University for isotope ratio determination.

This study was supported by the Geography Programs, Office of Naval Research, through the Coastal Studies Institute, Louisiana State University, under Contract N00014-69-A-0211-0003, Project NR 388 002.

REFERENCES

Allen, R. C., E. Garvish, G. M. Friedman, and J. E. Sanders, Arag-
onite cemented sandstone from the outer continental shelf off
Delaware Bay: submarine lithification mechanism yields product
resembling beachrock, *J. Sediment. Petrology, 39,* 136. 1969.

Emery, K. O., and D. Hoggan, Gases in marine sediments. *Amer.
Ass. Petrol. Geol. Bull., 42,* 2174, 1958.

Fredericks, A. D., and W. M. Sackett, Organic carbon in the Gulf of
Mexico, *J. Geophys. Res., 75,* 2199, 1970.

Garrison, R. E., L. J. Luternaur, E. V. Grill, R. D. McDonald, and
J. W. Murray, Early diagenetic cementation of Recent Sands, Fraser
River Delta, British Columbia, *Sedimentology, 12,* 27, 1969.

Ho, C. L., Seasonal changes in sediment and water chemistry in
Barataria Bay, *Louisiana State Univ., Coastal Studies Inst.,
Bull. 6,* 67, 1971.

Hood, M. A., and A. R. Colmer, Seasonal bacteria studies in Barataria
Bay, *Louisiana State Univ., Coastal Studies Inst., Bull. 6,*
16, 1971.

Jones, J. L., C. B. Leslie, and L. E. Barton, Acoustic characteristics of underwater bottoms, *J. Acoust. Soc. Amer., 36,* 154, 1964.

Koyama, T., Gaseous metabolism in lake sediments and paddy soils, in
Advances in Organic Geochemistry, edited by U. Colombo and G. D.
Hobson, pp. 363-375, McMillan, New York, 1964.

Levin, F. K., The seismic properties of Lake Maracaibo, *Geophysics,
27,* 35, 1962.

Nissenbaum, A., B. J. Presley, and I. R. Kaplan, Early diagenesis in
a reducing fjord, Saanich Inlet, British Columbia - I. Chemical
and isotopic changes in major components of interstitial water,
Geochim. et Cosmochim. Acta, 36, 1007, 1972.

Presley, B. J., and I. R. Kaplan, Changes in sulfate, calcium and
carbonate from interstitial water of nearshore sediments.
Geochim. et Cosmochim. Acta, 32, 1037, 1968.

Reeburgh, W. S., An improved interstitial water sampler, *Limnol.
Oceanogr., 12,* 163, 1967.

Reeburgh, W. S., Determination of gases in sediments, *Environ. Sci.
and Technol., 2,* 140, 1968.

Reeburgh, W. S., Observations of gases in Chesapeake Bay sediments,
Limnol. Oceanogr., 14, 368, 1969.

Swinnerton, J. W., and V. J. Linnenbaum, Determination of the C_1 to
C_4 hydrocarbons in sea water by gas chromatography, *J. Gas
Chromatogr., 5,* 570, 1967.

Takai, Y., and T. Kamura, The mechanism of reduction in water-logged paddy soil, *Folia Microbiol.*, *(Prague)*, *11*, 304, 1966.

Whelan, T., and H. H. Roberts, Carbon isotope composition of carbonate nodules from a freshwater swamp, *J. Sediment. Petrology*, *43*, 54, 1973.

HYDROCARBON GAS (METHANE) IN CANNED DEEP SEA DRILLING PROJECT CORE

SAMPLES

Richard D. McIver

Esso Production Research Company
Houston, Texas 77001

ABSTRACT

There were wide variations in the amounts of gas retained by Deep Sea Drilling Project cores at the time they were brought to the deck of the *Glomar Challenger*. To date, about 60 samples from 12 different sites, sealed in metal cans, have been analyzed for quantities and composition of gases. The quantities of gas varied from background, a few tens of parts per million by volume, to over 215,000 ppm by volume; virtually all the gas was methane. While the measured quantities are below those that could be dissolved in the pore water of the sediments, some of these values are very large when the time (i.e., for gas to escape in response to reduced pressures) between their removal from the subsea environment and their being sealed into the cans is considered.

INTRODUCTION

Quite often, Deep Sea Drilling Project (DSDP) personnel aboard the *Challenger* noted gas in the deep-sea muds when the core liners were removed from the core barrels. Sometimes the odor of gas was noted, but more often the gas revealed itself by the strong expansion of the sediment and the formation of gas gaps along the length of the core, together with the extrusion of the mud from both ends of the liner. Sometimes this expansion continued for an hour or two. The expansion gaps were large (e.g., up to 10 or more centimeters) in the meter-long segments from the Cariaco Trench, which were frozen directly in the liners. The nature and quantity of this gas have been of great interest to DSDP scientists. Shipboard analyses [*Claypool and Kaplan, 1974*] showed the gas to be primarily methane, but the quantities have only been described in such terms as 'very gassy', etc. Recently, anomalous seismic velocities in a 'gassy' section on the Blake Outer Ridge and a strong bottom-paralleling reflector, which cut bedding plane reflections, together with some circumstantial mineralogical evidence [*Lancelot and Ewing, 1972*], led to speculation, now widely accepted, that the gas in the sediment immediately below the sea floor was in the hydrate form. Moreover, the combined low temperatures and high pressures of the bathyal and abyssal environments are such that any hydrocarbon gases should be hydrated [*Katz, 1971*].

In order to find out more about the gas, the DSDP Advisory Panel on Organic Geochemistry recommended in 1971 that small portions of some of the cores, particularly those that appeared to be 'gassy', be sealed in metal cans as quickly as possible after their recovery. Subsequent analyses of the total trapped gases would then give the quantity of gas still remaining in the sediment at the time of canning. Admittedly, some (perhaps most) of the gas originally in the sediment would be lost during the trip to the *Challenger* deck and during subsequent handling, but the amounts might reveal the order-of-magnitude level of gases in place. Moreover, the results might reveal significant trends with depth or age, or contrasts from one area to another.

Collection of the canned samples began on Leg 18, off Oregon and Alaska. Subsequently, sporadic samples were received from Legs 19, 21, and 23. To date about 58 samples from 12 sites have been analyzed. These were generally from those sites where the sediments obviously contained gas. When the samples arrived, the cans were punctured through a rubber septum clamped to the side of the can. A measured volume of each can's air space was removed with a hypodermic syringe and introduced into the inlet system of a Perkin Elmer Model 154B gas chromatograph with a hydrogen-flame detector. [The column is a 2-meter P.E. column B (di-2-ethylhexylsebacate); temperature, 46°C; helium pressure, 8 psig; hydrogen pressure, 8 psig.] From the volumes of gas in the cans, the volume of the sediment, and the gas chromatographic results, the gas contents were calculated as parts-per-million of gas by volume in the sediment, (i.e., cm^3 of gas at STP per million cm^3 of sediment as received).

After the above analysis, which was sometimes run in duplicate, the can was opened, and a sample of each sediment was beaten vigorously in a Waring blender to remove any remaining gases. The additional quantities were so small (only a few parts per million) that this procedure was not continued after the first few analyses. Still another portion of each sample of sediment was subjected to an organic carbon determination by procedures reported previously [*Gehman, 1962*].

TABLE 1. Residual Gas Contents
Deep Sea Drilling Project Muds

Leg	Site	Location	Core No.	Section No. and Depth	Approx. Depth Meters	Organic Carbon (Percent)	Hydrocarbon Gas Content ppm by Vol.	Percent Methane
18	174A	Astoria Fan	6	3-btm.	80	0.21	20,700	100.0
		off Oregon	6	4-btm.	82	0.35	13,200	100.0
		2818m water	11	5-btm.	130	0.70	28,900	100.0
		depth (wd)	11	6-top	131	0.60	24,200	99.5
			13	2- ?	144	N/A	Trace	-
			15	4-top	166	0.37	24,500	99.5
			15	4-top	166	0.34	Trace	-
			19	4-btm.	205	0.23	36,000	100.0
			29	3- ?	297	N/A	22,900	99.8
			34	3-btm.	373	0.36	15,100	100.0
			34	3-btm.	373	0.32	Trace	-
			37	4- ?	509	N/A	1,100	98.7
			40	5-top	764	0.21	Trace	-
			40	5-top	764	0.16	41,400	99.5
18	176	Oregon Shelf	3	2-150cm	17	0.30	48,000	100.0
		wd-193m	3	2-150cm	17	0.38	76,000	100.0
			4	5- ?	21	N/A	71,100	100.0
			4	6-150cm	23	0.17	33,200	100.0
			5	6-150cm	40	0.35	36,600	100.0
			5	6-150cm	40	0.37	40,900	100.0
18	180	E. Aleutian	12	2- 5cm	149	0.16	2,100	98.7
		Trench	12	2- 5cm	149	0.18	1,700	98.3
		wd-4923m	15	3- ?	246	N/A	Trace	-
			18	1- 8cm	271	0.30	1,400	99.8
			18	1- 8cm	271	0.24	11,600	100.0
			18	4- ?	275	N/A	42,800	100.0
			20	5- ?	415	N/A	55,700	100.0
			24	3-150cm	454	0.43	1,500	99.9
			24	3-150cm	454	0.43	Trace	-
19	185	Bering Sea	20	2- 0cm	664	0.49	39,700	100.0
		Umnak Plateau	20	3-150cm	665	0.57	11,000	100.0
		wd-2110m	21	3- 0cm	674	0.45	101,800	100.0
19	186	Aleutian	3	2- 0cm	1	0.53	112,900	100.0
		Trench	3	3-150cm	3	0.46	17,100	100.0
		North Wall	3	5- 0cm	6	0.86	40,800	100.0
		wd-4552m	3	5-150cm	6	0.68	3,400	100.0
			3	6-150cm	15	0.74	27,600	100.0
19	186	Aleutian	3	6-150cm	15	0.62	216,700	100.0
		Trench	4	5-150cm	55	0.50	49,900	100.0
		North Wall	6	1-150cm	107	0.36	26,800	100.0
		wd-4552m	8	1- 0cm	132	0.38	1,400	91.0
			9	7-btm.	167	0.56	28,900	100.0
			20	5- 0cm	465	0.40	27,800	100.0
19	189	Bering Sea	5	4- 0cm	152	0.26	Trace	-
		wd-3437m	6	1-150cm	212	0.47	19,900	100.0
19	191	Bering Sea	4	5-150cm	84	0.66	119,000	100.0
		wd-3854m	4	6-150cm	86	0.71	48,000	100.0
			5	5-150cm	140	0.40	107,100	100.0
21	204	E. of Tonga	5	4- 0cm	99	0.29	Trace	-
		Trench wd5354m	5	4- 0cm	99	0.40	Trace	-

TABLE 1.   (Continued)

Leg	Site	Location	Core No.	Section No. and Depth	Approx. Depth Meters	Organic Carbon (Percent)	Hydrocarbon Gas Content ppm by Vol.	Percent Methane
21	210	Coral Sea wd-4643m	5	1- 0cm	36	0.12	Trace	-
			5	2- 0cm	37	0.14	Trace	-
23	222	Arabian Sea wd-3546m	5	4- ?	132	0.28	87,900	100.0
23	229	S. Red Sea wd-852m	2	3- 0cm	50	0.94	84,500	100.0
			2	4- 50cm	52	1.14	165,700	100.0
23	229A	S. Red Sea wd-852m	3	4- 0cm	52	1.01	32,600	100.0
			12	4- ?	153	0.67	1,000	99.1
			13	6-150cm	166	1.18	Trace	-

RESULTS

The results of analyses are listed in Table 1, along with locations and identification of the samples. In addition, in order
to make it easier to visualize the range of values and their distribution, the data are also presented in histogram form in Figures
1 and 2. Figure 1 shows the distribution of residual gas contents,
and Figure 2 shows the distribution of values for residual gas contents per unit percentage of organic carbon.

As the table shows, the gas contents are erratic. There seem
to be no trends with depth, nor any strong correlation with the organic carbon contents. Of course, not much significance can be attached to the variations, for they simply reflect the amount remaining in the sediment when the can was sealed. A coarse-grained (i.e.,
permeable and porous) sediment would release its gas much more rapidly than a fine-grained one. Therefore, the higher values are
probably more indicative of the overall gas contents of the sediment
column. And considering the losses in handling, even the highest
values must be considered minimal.

The more narrow and uniformly higher range of values in site
176 are puzzling. It may be due to more uniform and finer grain
size of the sediments. Still the ranges of values of the other two
sites overlap these values.

In the last column, it appears that some of the hydrocarbon
gases may be as much as 9% ethane and higher homologs, but the lower values are associated with the lower total hydrocarbon gas contents. In these cases, the trace of ethane, which occurs in

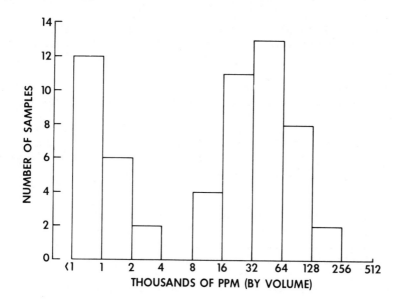

Fig. 1. Residual hydrocarbon-gas contents of canned DSDP cores

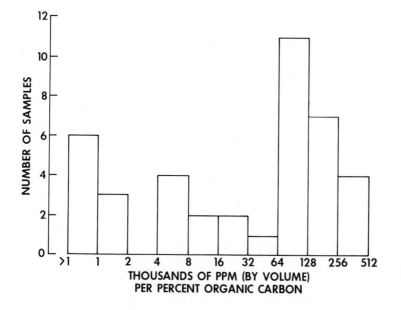

Fig. 2. Residual hydrocarbon-gas contents as a function of organic carbon in canned DSDP cores

virtually every sample, sometimes appears to be a significant percentage of the gas even though it is only slightly higher than the background level. Probably not one of the values below 99.0% is significantly different than those above. So, for all intents and purposes, we can consider these gases to be virtually all methane.

This predominance of methane in the hydrocarbon gases suggests that the gas is produced microbially in the near-bottom sediments [*Claypool and Kaplan, 1974*]. Some carbon isotope analyses were run on gases from the sediments on the same cruise and same sites, and the methane was found to have carbon isotope ratios as high as -80% versus PDB. This seems to confirm the microbial origin of the methane, but further work needs to be done in this area.

Lower values of gases, especially just trace amounts, may be due to low gas contents or gas generating ability of the sediments themselves such as those from sites 204 and 210. In the other cases, it is more likely that these are artifacts from longer sample exposure to the atmosphere before the canning step, or of poor seals or leaks in the cans themselves. A few of the cans show evidence of leakage, i.e., salt crystals around tiny spots on the shipboard sealed rim. These samples were not run for residual gas contents. Moreover, it is entirely possible that other leaking cans, without visible manifestations of a leak, were submitted for analyses.

The data presented in Figure 1 suggest that residual gas contents are lognormally and bimodally distributed among deep-sea sediments. The distribution of ratios of residual gas to content of organic carbon (Figure 2) suggests that some types of organic matter do not produce as much gas as others. The elongated end of this skewed distribution may be due to causes relating to gas origin, but they may also reflect loss of most of the gas from the containers before analyses, as discussed above. Nevertheless, the data from the 58 residual gas determinations serve as a basis for comparing and contrasting results from individual sites or samples.

Site 185 was one of two chosen for study of a 'bottom-simulating reflector' that has been attributed to gas-hydrate zones in near-bottom, deep-sea sediments. A major gassy zone was penetrated bracketing the 670-meter depth of the reflector, and an attempt was made to collect some cores at *in situ* pressures [*Bryan, 1974*]. This pressure-coring failed. Three samples were taken from the regular cores, sealed in cans and sent to us. These three samples had moderate to high residual gas contents. The one from beneath the 'reflector' had over 100,000 ($\simeq$ 400,000 ppm) at this depth (pressure) and estimated temperature in the sediment's interstitial water. Still, it is a large quantity of gas when the coring and handling procedures are considered.

The residual gas contents of the Site 186 samples were erratic, but they did include the most gas-rich sample run in this program to date. The 216,700 ppmv methane gas in core 3, section 5, is only about a third of the gas concentration theoretically soluble in the interstitial water of this sediment. This is remarkable retention considering the coring and handling procedures, unless the gas was tied up as the hydrate, which decomposed slowly after removing the core from its original setting.

Although little can be said about the distribution and range of values of hydrocarbon gases in all deep-sea sediments based on this work alone, the amounts are significant, and the data presented herein serve as values against which later samples can be compared. Undoubtedly, hydrocarbon gas is a common constituent of sediments under the oceans; it will be interesting to see what future samples contain.

ACKNOWLEDGMENTS

The author is indebted to Harvey M. Fry, Jim Morgan, Shirley Tillotson, Tom Conary, and Rosie Ramirez who performed the analyses on which this report is based; to the National Science Foundation and the Deep Sea Drilling Project which made the samples available; and to Esso Production Research Company for its continuing support for this work.

REFERENCES

Bryan, G. M., *In situ* indications of gas hydrate, in *Natural Gases in Marine Sediments*, edited by I. R. Kaplan, pp. 299-308, Plenum Press, New York, 1974.

Claypool, G. E., and I. R. Kaplan, The origin and distribution of methane in marine sediments, in *Natural Gases in Marine Sediments*, edited by I. R. Kaplan, pp. 99-139, Plenum Press, New York, 1974.

Gehman, H. M., Organic matter in limestones, *Geochim. et Cosmochim. Acta, 26*, 885, 1962.

Katz, D. L., Depths to which frozen gas fields (gas hydrates) may be expected, *J. Petr. Tech., 23*, 419, 1971.

Lancelot, Y., and J. I. Ewing, *Initial Reports of the Deep Sea Drilling Project*, vol. 11, pp. 791-800, U. S. Government Printing Office, Washington, 1972.

DISSOLVED GASES IN CARIACO TRENCH SEDIMENTS: ANAEROBIC DIAGENESIS

Douglas E. Hammond

Lamont-Doherty Geological Observatory
Columbia University
Palisades, New York 10964

ABSTRACT

Gas pockets sampled in DSDP cores retrieved from the Cariaco Trench are about 90% CH_4 and 8% CO_2, with the remainder consisting of H_2O, N_2, and C_2H_6. Conversion of this data to *in situ* dissolved gas concentrations is hindered by the lack of an *in situ* sampler or a suitable internal standard, but the best estimate for the interstitial water dissolved gas concentrations is $[CH_4] \approx 20$ mmol/l, $[N_2] \approx 0.07$ mmol/l, $[C_2H_6] \approx 4$ μmol/l. The carbon in CH_4 is about 1% of carbon in organic compounds in the sediment, and is sufficiently abundant to suggest that CO_2 reduction to CH_4 causes carbonate precipitation.

Below 80 meters in the sediment, $[N_2]$ is present at less than 20% of its concentration in sea water. This is attributed to microbial activity. Calculation of pe from the pairs HCO_3^-/CH_4, N_2/NH_4^+, and C_2H_6/CH_4 suggests that CO_2 reduction to CH_4 and N_2 reduction to NH_4^+ are accomplished by different microorganisms and C_2H_6 is a residual product of the degradation of larger molecules.

INTRODUCTION

The Cariaco Trench is a basin about 150 miles long and 50 miles wide off the coast of Venezuela (Figure 1). It reaches a maximum depth of about 1500 meters, but is surrounded by a sill at 200 meters that restricts exchange with the deep ocean. As a result, the temperature and salinity are nearly uniform at 16.9°C and 36.6°/₀₀ below 200 meters. High productivity in the surface waters and a large influx of continental detritus have combined with this restricted

71

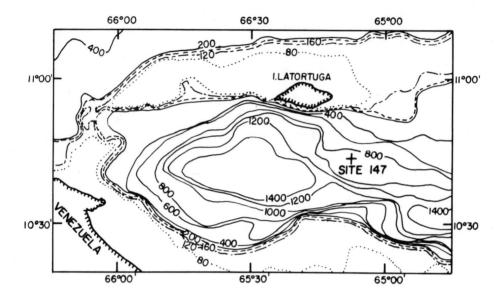

Fig. 1. Bathymetry of the Cariaco Trench and the location of DSDP Site 147 (+) [after *Richards and Vaccaro, 1956*]. Depth is in meters.

circulation to create anoxic conditions in the deep water. $[NO_3^-]$ and $[O_2]$ are absent below 400 meters, and $[H_2S]$ is present [*Richards, 1960*].

On Leg 15 of the Deep Sea Drilling Project, a site was drilled in the Cariaco Trench in 882 meters of water for intensive geochemical sampling (see *Edgar et al.* [*1973*] for the initial results of other programs). The sediment is rather uniformly a calcareous clay containing authigenic pyrite and dolomite and a high fraction of organic carbon, about 2% by dry weight. The sedimentation rate was about 50 cm per thousand years over the 176 meters penetrated [*Edgar et al. 1973*]. Interstitial water analyses by *Gieskes* [*1973*]; *Presley et al.* [*1973*]; and *Sayles et al.* [*1973*] showed that $[SO_4^=]$ disappears in the upper 5 meters of the sediment. Associated with this are decreases in $[Ca^{++}]$ and $[Mg^{++}]$, and an increase in alkalinity, apparently reflecting the oxidation of organic matter by sulfate reduction and the subsequent precipitation of pyrite and carbonate. Below 40 meters in the sediment, $[Ca^{++}]$ and $[Mg^{++}]$ become nearly constant with depth, but alkalinity rises from about 12 meq/kg to a maximum of 34 meq/kg at 100 meters, and then decreases to 24 meq/kg at 176 meters.

SAMPLE COLLECTION AND ANALYSIS

Cores retrieved from below 40 meters in the sediment contained pockets of gas when they arrived on deck. Using a device designed by Ross Horowitz and Taro Takahashi [see *Horowitz et al. 1973*]. ten of these gas pockets were sampled. Briefly, the procedure was to clamp a steel block around the plastic core liner and evacuate the volume between the liner and the block through a stainless steel needle that is screwed into the block. The needle is then driven through the core liner, the pocket pressure is measured, and an aliquot of gas is collected in an evacuated flask. These flasks were returned to Lamont-Doherty where an aliquot was taken for gas chromatographic analysis [*Hammond et al. 1973*].

This analysis showed the gas pockets to be primarily CH_4 and CO_2, and to lie within the compositional range of many natural gases. Small amounts of O_2 were observed and were attributed to air contamination during sampling and handling; and the N_2 was corrected for this. The results listed in Table 1, calculated on an air-free basis, show that the CO_2 fraction rises with depth and then decreases slightly, with the N_2 fraction decreasing to nearly zero. As the total pressure in the pocket was measured, Table 1 can be used to calculate the partial pressure of each gas.

One question which must be considered is whether gas pockets are in equilibrium with the interstitial water. Two types of *p*H measurements were made as part of the sampling program: one by inserting electrodes directly into the sediment (punch-in), and one on water squeezed from the sediment by an hydraulic press at either 4°C or 22°C (cold squeeze, CS, and warm squeeze, WS). The results of these measurements, when recalculated to 25°C, are plotted in Figure 2. The *p*H, calculated from the gas pocket p_{CO_2} (Figure 3) and the alkalinity profile determined on interstitial water [*Gieskes, 1973; Hammond, 1973*] is also plotted in Figure 2. As squeezed samples were observed to effervesce during collection, CS and WS *p*H values are probably too high. The consistency between punch-in *p*H measurements and those calculated from p_{CO_2} suggests that both numbers are reasonable (within 0.2 *p*H units), and that the gas pockets are probably not far from being in equilibrium with the interstitial water, as the kinetics of CO_2 equilibration will be much slower than those of non-reactive gases.

The methane partial pressure profile (Figure 4) is nearly the same as the total pressure profile. Although some increase in p_{CH_4} might be expected with depth, where more time has elapsed to generate methane, no such correlation is apparent. The profile probably reflects the confining pressure which is exerted by the soft sediment plugs as they seal the gas pocket in the core liner, rather than indicating gas content. Thus, a gas pocket should expand until

TABLE 1. Gas Pocket Composition*

| Sample | Depth (m) | Total Pocket Pressure (atm) | Air Contamination (%) | Normalized to 100% | | | | | | N_2 Error as $\Delta N_2/N_2$ (%) |
				CH_4 (%)	CO_2 (%)	H_2O (%)	C_2H_6 (%)	N_2 (%)	
147-6-4	47	1.87	5.7	95.4	1.16	0.9	0.004	2.45	7
147-B-6-2	51	0.96	3.1	92.5	4.61	1.6	0.023	1.17	8
147B-7-4	63	2.21	3.5	92.7	5.57	1.1	0.012	0.53	30
147-10-3	82	1.36	3.1	91.3	6.84	1.5	0.015	0.35	29
147B-9-4†	83	1.12	3.4	86.1	13.5	–	–	0.44	57
147B-11-3	100	2.08	3.5	85.8	12.3	1.8	0.014	0.16	63
147-14-3†	118	1.09	3.3	87.8	11.9	–	0.020	0.27	32
147C-2-2	128	1.09	4.8	85.9	12.8	1.1	0.025	0.16	100
147C-4-4	148	1.50	3.3	87.5	10.6	1.4	0.034	0.42	23
147C-7-3	175	1.29	2.6	91.8	6.70	1.4	0.047	0.07	170
Median		1.33	3.4	90	8.7	1.4	0.02	0.3	
Error as random	5			2	3	20	30		
Δgas/gas (%) systematic				2	3	10	20		

*Hammond et al. [1973].

†Calculated on a water-free basis.

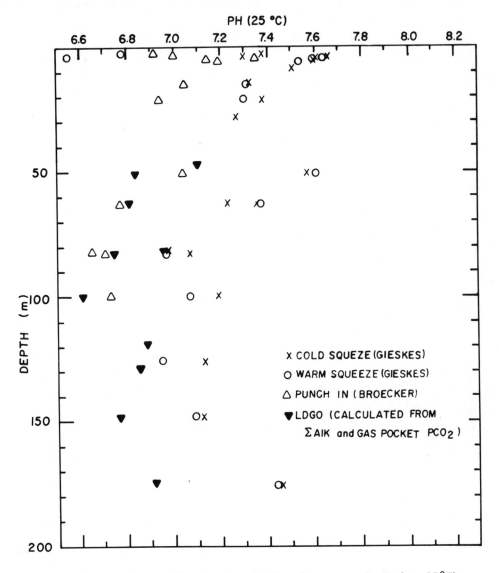

Fig. 2. Comparison of calculated pH and measured pH (at 25°C).

its pressure decreases sufficiently for friction at the sediment-core liner interface to confine it.

The p_{N_2} profile (Figure 5) shows a rapid decrease with depth. One of two explanations could account for this: N_2 is consumed by bacteria *in situ*; N_2 is conservative and the decrease represents only dilution with CH_4.

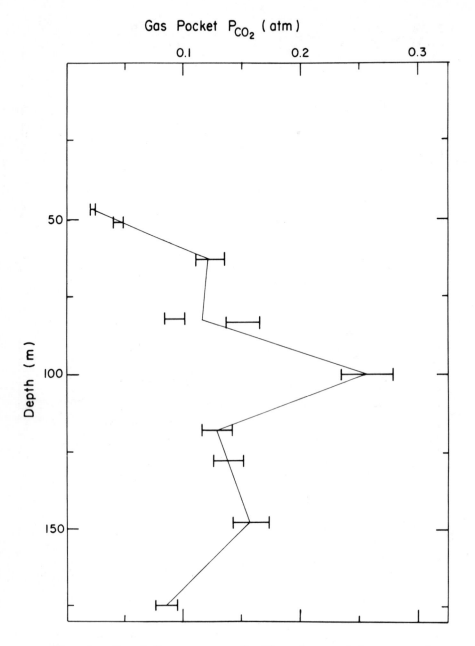

Fig. 3. Partial pressure of CO_2 observed in gas pockets.

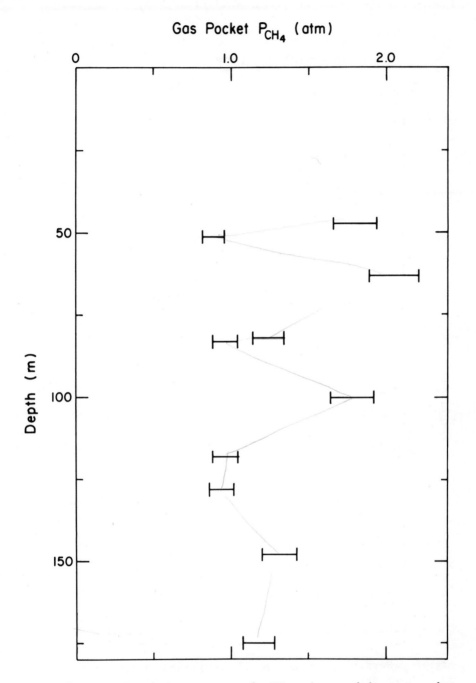

Fig. 4. Partial pressure of CH_4 observed in gas pockets.

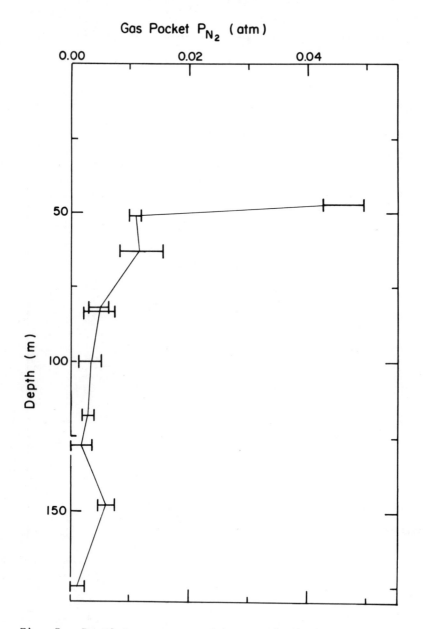

Fig. 5. Partial pressure of N_2 observed in gas pockets.

When interstitial water is trapped by sediment deposition, it should contain dissolved $[N_2]$ close to that contained by sea water, with similar T-S characteristics that has been equilibrated with the atmosphere, or 0.44 mmol/1 (from the solubility tables of *Weiss* [*1970*] for T = 17°C, S = 36.6°/$_{oo}$). The temperature in this basin cannot have varied much from 17° during the last 400,000 years, as oxygen isotope studies of foraminifera indicate temperature variations of only a few degrees in the surface waters of the Caribbean Sea over this interval [*Broecker and van Donk, 1970*]. The solubility of N_2 is rather low, and when sampled, most of the N_2 present *in situ* in the interstitial water should be in the gas pocket. Throughout the core, the porosity is nearly uniform at 70%. Since estimates indicated that gas pockets occupied about 20% of the volume of the core at the time of sampling, interstitial water occupied 0.8 x porosity = 56% of the core, and the *in situ* $[N_2]$ can be estimated

$$[N_2] \simeq p_{N_2} \left(\frac{V_{gas}}{RTV_{water}} \right)$$

where R = ideal gas constant and T = temperature. Below 80 meters, $p_{N_2} \simeq 5 \times 10^{-3}$ atm (Figure 5), indicating that $[N_2] \simeq$ 0.08 mmol/1 *in situ*, less than 20% of the anticipated 0.44 mmol/1. The remainder must either have been consumed by microorganisms *in situ* or escaped during core retrieval.

First, the hypothesis that $[N_2]$ is consumed will be discussed. *Thorstenson* [*1970*] has shown that in environments sufficiently reducing to allow methane to be present in significant quantities, the stable form of nitrogen is NH_4^+. Although N_2 reduction has not been observed previously in marine sediments, *Zelitch* [*1951*] has shown that bacteria of the genus *Clostridium* are known to convert N_2 to NH_3, and in particular, *Cl. pasteurianum* has been shown to utilize N_2 for protein synthesis in laboratory experiments, even in the presence of NH_3. Thus, it should not be surprising to find that N_2 is not conservative over long time spans in highly reducing environments. Apparently, amino acid degradation masks any possible reaction, as $[NH_4^+]$ is observed to increase from 4 mmol/1 at 63 meters to 6 mmol/1 at 82 meters [*Gieskes, 1973*], over twice the 0.9 mmol/1 increase that could be produced by N_2 reduction.

The second possible explanation for the p_{N_2} decrease, that N_2 is conservative and extensive gas loss has occurred, cannot be definitively ruled out, but, if true, leads to inconsistencies, as follows: If N_2 were conservative, it could be used as an internal tracer. Assuming $[N_2]$ in the interstitial waters *in situ* to be the same as in the deep waters, $[CH_4]$ *in situ* can be calculated from the CH_4/N_2 ratio observed in the gas pocket (see *Hammond et al.* [*1973*] for details). The large errors in determining the gas pocket N_2 fraction limit the accuracy of this calculation, but allow determination of at least lower limits for methane abundance *in situ* (Figure 6). This approach indicates rather high *in situ* concentrations at depth. Using data obtained by *Culberson and McKetta* [*1951*] for the solubility of methane in distilled water, and making a 20% correction for the effect of salinity, a bubble point curve (dashed line) can be calculated. This line represents the minimum methane concentration necessary to form bubbles *in situ*. Most values below 80 meters exceed this, indicating that bubbles should exist *in situ*. However, if bubbles actually were to form, they should migrate upward until they dissolve. Therefore, the *in situ* profile should rise until it reaches the bubble point curve. As the pressure is reduced from 100 atm *in situ* to 1 atm, the gas volume released from a sample lying on this curve should occupy 70% of the core volume rather than the observed 20%. Another argument against extensive gas loss is the observation that the pressure in the pockets was often greater than 1 atm, suggesting that the sediment plugs make reasonably efficient traps for small volumes of gas. These points suggest that extensive gas loss is unlikely.

It is important to point out that the unusually high temperature in these sediments should prevent clathrate formation. The locus of the P-T curve in this environment falls about 5 to 6°C above the stability field for clathrates determined by *Deaton and Frost* [*1946*] when salinity corrections are made.

As $[N_2]$ is apparently non-conservative, the best method of converting gas pocket partial pressures into *in situ* concentrations is to use the visual estimate of gas pocket fraction mentioned earlier. This approach lacks resolution of gas content and has the unfortunate consequence of suggesting that $[CH_4]$ is approximately uniform with depth at about 20 mmol/l (Table 2), which is probably not the case. This corresponds to an *in situ* p_{CH_4} of about 25 atm, well below the *in situ* value of 100 atm. The $[N_2]$ decreases with depth and $[C_2H_6]$ increases with depth should be correct.

DISCUSSION

Having established the concentrations of dissolved gases *in situ*, probably within a factor of 2, the data can be used to examine biologically catalyzed reactions associated with anaerobic diagenesis. It is interesting to observe that large amounts of methane are not found until well below the sulfate reduction zone, as *Mechalas* [*1974*] would predict on microbiological grounds. It seems rather clear that in this environment much of the methane is produced by the reduction of CO_2 by the addition of hydrogen through the efforts of microorganisms [see *Claypool and Kaplan; Lyon; Mechalas, 1974*].

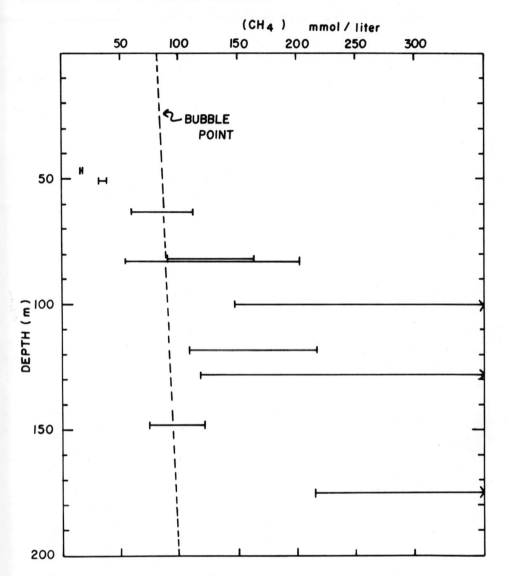

Fig. 6. *In situ* methane concentrations assuming that N_2 is conservative. The dashed line represents the concentration necessary to form bubbles *in situ*.

The source of the electrons to do this is ultimately the organic matter in the sediment. Although the pathways in this environment are unknown, one possibility is that H_2 is split from the organic matter by one group of microorganisms and is then

TABLE 2. Best Estimate of *in situ* Concentrations

Sample	Depth (m)	$[CH_4]^*$ mmol/l	$[HCO_3^-]^\dagger$ mmol/l	$[N_2]^*$ mmol/l	$[NH_4^+]^\dagger$ mmol/l	$[C_2H_6]^*$ µmol/l	$pH^\S$
147-6-4	47	28	10.4	0.70	3.25	1.	7.14
147B-6-2	51	14	10.4	0.17	3.30	3.5	6.84
147B-7-4	63	32	23.4	0.18	4.15	4.3	6.81
147-10-3	82	20	29.0	0.07	6.10	3.2	6.96
147B-9-4	83	15	29.0	0.07	6.15	-	6.75
147B-11-13	100	28	34.8	0.05	6.85	4.5	6.61
147-14-3	118	15	34.0	0.05	6.50	3.5	6.89
147C-2-2	128	14	33.4	0.03	6.20	4.1	6.85
147C-4-4	148	21	31.6	0.09	9.60	8.2	6.77
147C-7-3	175	19	24.2	0.01	8.15	9.7	6.92
Cariaco Trench							
Deep water		$0.007^\ddagger$		0.44		$0.003^\ddagger$	

* *Hammond et al.* [*1973*].

$\dagger$ Interpolated from *Gieskes* [*1973*]; assuming $[HCO_3^-] \approx$ Alkalinity.

$\ddagger$ *Lamontagne et al.* [*1972*].

$\S$ *Hammond* [*1973*]; calculated from P_{CO_2} and Alkalinity.

oxidized by methane bacteria. *Zobell [1947]*, who carried out laboratory studies on marine sediments, indicates that H_2 evolution is quite possible. $[N_2]$ may have been reduced by a similar mechanism.

Nissenbaum et al. [1972] have suggested that CO_2 reduction can cause carbonate precipitation by $2HCO_3^- + 4H_2 + Ca^{++} \rightarrow CH_3 + CaCO_3 + 3H_2O$. Sufficient CH_4, 20 mmol/l, is present to suggest that this reaction could have caused the 10 meq/l decrease in alkalinity observed by *Gieskes [1973]* between 118 meters and 175 meters. Organic carbon is about 2% of the sediment (dry weight), the porosity is about 70%, and taking the density of dry sediment as 2.5 gm/cm^3 indicates that the amount of carbon in CH_4 is about 1% of that in organic material. Degradation of the organic matter should be far from complete.

Although many of the reactions occurring in the sediment are biologically catalyzed, it is instructive to approach the system from a thermodynamic viewpoint. A series of oxidation reactions can be written involving electron transfer among the dissolved gases considered here

$$1) \quad CH_4 + 3H_2O \rightarrow HCO_3^- + 9H^+ + 8e^-$$

$$2) \quad 2NH_4^+ \rightarrow N_2 + 8H^+ + 6e^-$$

$$3) \quad 2CH_4 \rightarrow C_2H_6 + 2H^+ + 2e^-$$

The equilibrium constant for reaction (1) is

$$K_1 = \frac{\left(a_{H^+}\right)^9 \left(aHCO_3\right) \left(a_e\right)^8}{\left(a_{CH_4}\right) \left(a_{H_2O}\right)^3}$$

and

$$\Delta G_1 = -RT \ln K$$

where a_e is defined as the activity of electrons in the system, R is the ideal gas constant, T is the absolute temperature, and ΔG_1 is the free energy change for reaction (1). Defining $pe = -\log a_e$,

$$pe_1 = \frac{\Delta G_1}{8 \times 2.3RT} - 1.125pH + 0.125 \left[\log \frac{\gamma_{HCO_3^-}}{\gamma_{CH_4} a_{(H_2O)}^3} + \log \frac{[HCO_3^-]}{[CH_4]} \right]$$

Similarly,

$$pe_2 = \frac{\Delta G_2}{6 \times 2.3RT} - 1.33pH + 0.167 \left[\log \frac{\gamma_{N_2}}{(NH_4^+)^2} + \log \frac{[N_2]}{[NH_4^+]^2} \right]$$

$$pe_3 = \frac{\Delta G_3}{2 \times 2.3RT} - pH + 0.5 \left[\log \frac{\gamma_{C_2H_6}}{\gamma_{(CH_4)}^2} + \log \frac{[C_2H_6]}{[CH_4]^2} \right]$$

where γ_i activity coefficient of species i. If the species in Table 2 are in thermodynamic equilibrium, the three values of pe calculated for each sample should be identical (see *Stumm and Morgan* [1970] for a more detailed treatment of the theory). Choosing activity coefficients from *Garrels and Thompson* [1962] and calculating ΔG at 100 meters in the sediment (setting $P \simeq 105$ atm and $T \simeq 25°C$)

$$\Delta G_i = \Delta G_i^° (1 \text{ atm, } 25°C) + (P - 1) \times \Delta V_i$$

from free energy data tabulated by *Berner* [1971] and partial molar volumes determined by *Owen and Brinkley* [1941] and *Kobayashi and Katz* [1953], the relations become

$$pe_1 = 3.47 - 1.125pH + 0.125 \log \frac{[HCO_3^-]}{[CH_4]}$$

$$pe_2 = 5.23 - 1.333pH + 0.167 \log \frac{[N_2]}{[NH_4^+]^2}$$

$$pe_3 = 4.60 - 1.0pH + 0.50 \log \frac{[C_2H_6]}{[CH_4]^2}$$

The effect of pressure on ΔG for these reactions is rather small, and in making this calculation, the partial molar volume of electrons was arbitrarily chosen as zero.

Using the values in Table 2 (assuming $HCO_3^- \simeq Alk$ in this pH range), the points in Figure 7 can be obtained. Little depth dependence is apparent, and each reaction appears to characterize a distinct value of pe. Much of the scatter with depth for any one reaction may be attributed to a 0.2 unit uncertainty in pH, but this should introduce systematic errors in the pe values for all three reactions at any horizon. This effect can be partially eliminated by plotting the pe values for one reaction against the others as has been done by *Thorstenson* [1970]. If either reaction (2) or (3) is in equilibrium with reaction (1), the points should fall on the solid line in Figure 8. The sensitivity of the calculation to the data in Table 2 is also illustrated. If the concentrations of $[N_2]$, $[CH_4]$, and $[C_2H_6]$ are all an order of magnitude low, and the reactions are in equilibrium, the points should scatter about the dashed line. If $[N_2]$ alone is a factor of 10 too large, the triangles will all be moved along the vector labeled 'A'. If $[C_2H_6]$ is a factor of 10 too large, all circles will move along the vector 'B'. Thus Figure 8 offers little support for arguing that the values in Table 2 are unreasonable. It also suggests that $[N_2]$ and $[C_2H_6]$ are about 2 orders of magnitude more abundant than thermodynamics would predict, although below 80 meters the error in determining $[N_2]$ is rather large compared to the concentration, and the difference between pe_1 and pe_2 may not be significant.

It seems clear that the dissolved gases are not in thermodynamic equilibrium. This observation offers some insight into the microbiology occurring in the sediment. Apparently CO_2 and N_2 reductions must be undertaken by different microenvironments. As both processes occur in the same regions, apparently both types of organisms prefer similar environments. Ethane, far too abundant according to thermodynamics, may appear as a by-product of the degradation of larger organic molecules.

CONCLUSIONS

1) The concentration profile of dissolved gases with depth in Cariaco Trench sediment cannot be well characterized until an *in situ* sampler can be devised, or a measurement technique developed using an internal tracer such as A^{40}. Without this information, reliable models predicting the fate of carbon during diagenesis and the rate of methane production cannot be developed.

2) *In situ* $[CH_4]$ concentration in the interstitial water is about 20 mmol/l below 40 meters in Cariaco Trench sediment.

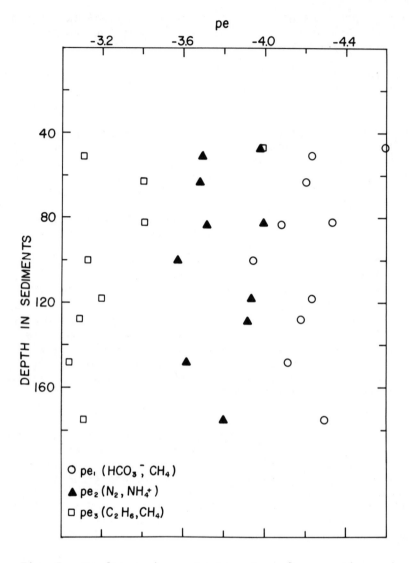

Fig. 7. *pe* for various electron transfer reactions.

This corresponds to only about 1% of the carbon in the organic reservoir, but is sufficient to suggest that CO_2 reduction causes carbonate precipitation, accounting for the alkalinity decrease observed by *Gieskes* [*1973*].

3) $[N_2]$ does not appear to be conservative in highly reducing sediments.

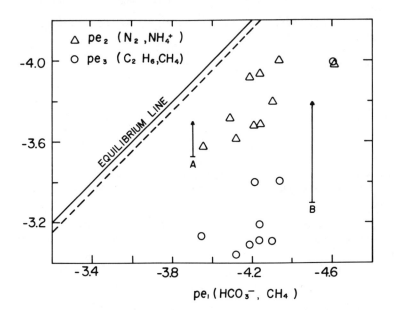

Fig. 8. pe_2 versus pe_1 and pe_3 versus pe_1. The solid line is the locus of points for reactions in equilibrium. If $[N_2]$, $[CH_4]$, and $[C_2H_6]$ in Table 2 were all a factor of 10 low, they would fall on the dashed line. 'A' is the vector along which all triangles would move if $[N_2]$ was lowered by a factor of 10, and 'B' is the vector along which circles would move if $[C_2H_6]$ were lowered by a factor of 10.

4) Thermodynamic considerations suggest that methane production and nitrogen reduction are accomplished by different microorganisms who enjoy the same environment, and that ethane is produced as a by-product of the breakdown of larger organic molecules.

ACKNOWLEDGMENTS

This paper exists as a result of the efforts of many people, especially Dr. N.T. Edgar and the JOIDES Geochemistry Panel who planned and executed the sampling program. Dr. Joris Gieskes kindly allowed me to quote his data prior to publication. Discussion with Drs. I. Kaplan and B. Mechalas was quite helpful, and Dr. Y-H. Li and Mr. T. Torgersen kindly reviewed the manuscript at Lamont-Doherty.

Marylou Zickl typed the manuscript, and Diann Warner did the drafting. But without the suggestions and guidance of Dr. Wallace Broecker none of this would have happened.

Financial support for the data analysis in this work has been provided by Atomic Energy Commission grants AT(11-1)2185 and AT(11-1)3180.

Lamont-Doherty Geological Observatory Contribution 2058.

REFERENCES

Berner, R. A., *Principles of Chemical Sedimentology*, McGraw-Hill New York, 1971.

Broecker, W. S., and J. van Donk, Insolation changes, ice volumes, and the O^{18} record in deep-sea cores, *Rev. of Geophys. and Space Phys.*, *8*, 169, 1970.

Claypool, G. E., and I. R. Kaplan, The origin and distribution of methane in marine sediments, in *Natural Gases in Marine Sediments*, edited by I. R. Kaplan, pp. 99-139, Plenum Press, New York, 1974.

Culberson, O. L., and J. J. McKetta, Solubility of methane in water at pressures to 10,000 psia, *Trans. AIME*, *192*, 223, 1951.

Deaton, W. M., and E. M. Frost, Jr., Gas hydrates and their relation to the operation of natural-gas pipe lines, *Bur. of Mines Monograph*, *8*, A. G. A., New York, 1946.

Edgar, N. T., J. B. Saunders, *et al.*, *Initial Reports of the Deep Sea Drilling Project*, vol. 15, pp. 169-216, U. S. Government Printing Office, Washington, 1973.

Garrels, R. M., and M. E. Thompson, A chemical model for sea water at 25°C and one atmosphere total pressure, *Amer. J. Sci.*, *360*, 57, 1962.

Gieskes, J. M., *Initial Reports of the Deep Sea Drilling Project*, vol. 20, pp. 813-830, U. S. Government Printing Office, Washington, 1973.

Hammond, D. E., *Initial Reports of the Deep Sea Drilling Project*, vol. 20, pp. 831-850, U. S. Government Printing Office, Washington, 1973.

Hammond, D. E., R. Horowitz, W. Broecker, and R. Bopp, *Initial Reports of the Deep Sea Drilling Project*, vol. 20, pp. 765-772, U. S. Government Printing Office, Washington, 1973.

Horowitz, R. M., L. S. Waterman, and W. S. Broecker, *Initial Reports of the Deep Sea Drilling Project*, vol. 20, pp. 757-764, U. S. Government Printing Office, Washington, 1973.

Kobayashi, R., and D. L. Katz, Vapor-liquid equilibria for binary hydrocarbon-water systems, *Ind. Eng. Chem.*, *45*, 440, 1953.

Lamontagne, R. A., J. W. Swinnerton, and V. J. Linnenbom, Comparison of C_1-C_3 hydrocarbons in various ocean environments (abstract), *Trans. Amer. Geophys. Union*, *53*, 405, 1972.

Lyon, G. L., Isotopic analysis of gas from the Cariaco Trench sediments, in *Natural Gases in Marine Sediments*, edited by I. R. Kaplan, pp. 91-97, Plenum Press, New York, 1974.

Mechalas, B. J., Pathways and environmental requirements for biogenic gas production in the ocean, in *Natural Gases in Marine Sediments*, edited by I. R. Kaplan, pp. 11-25, Plenum Press, New York, 1974.

Nissenbaum, A., B. J. Presley, and I. R. Kaplan, Early diagenesis in a reducing fjord, Saanich Inlet, British Columbia - I. Chemical and isotopic changes in major components of interstitial water, *Geochim. et Cosmochim. Acta, 36,* 1007, 1972.

Owen, B. B., and S. R. Brinkley, Calculation of the effect of pressure upon ionic equilibria in pure water and salt solutions, *Chem. Revs., 29,* 461, 1941.

Presley, B. J., J. Culp, C. Petrowski, and I. R. Kaplan, *Initial Reports of the Deep Sea Drilling Project,* vol. 20, pp. 805-180, U. S. Government Printing Office, Washington, 1973.

Richards, F. A., Some chemical and hydrographic observations along the north coast of South America - I. Cabo, Tres Puntas to Curacao, including the Cariaco Trench and the Gulf of Cariaco, *Deep-Sea Res., 7,* 163, 1960.

Richards, F. A., and R. F. Vaccaro, The Cariaco Trench, an anaerobic basin in the Caribbean Sea, *Deep-Sea Res., 3,* 214, 1956.

Sayles, F. L., F. T. Manheim, and L. S. Waterman, *Initial Reports of the Deep Sea Drilling Project,* vol. 20, pp. 783-804, U. S. Government Printing Office, Washington, 1973.

Stumm, W., and J. J. Morgan, *Aquatic Chemistry,* Wiley-Interscience, New York, 1970.

Thorstenson, D. C., Equilibrium distribution of small organic molecules in natural waters, *Geochim. et Cosmochim. Acta, 34,* 745, 1970.

Weiss, R. F., The solubility of nitrogen, oxygen, and argon in water and sea water, *Deep-Sea Res., 17,* 721, 1970.

Zelitch, I., Simultaneous use of molecular nitrogen and ammonia by *Clostridium pasteurianum, Proc. Nat. Acad. Sci., 37,* 559, 1951.

Zobell, C. E., Microbial transformation of molecular hydrogen in marine sediments, with particular reference to petroleum, *Amer. Ass. Petrol. Geol. Bull., 31,* 1709, 1947.

ISOTOPIC ANALYSIS OF GAS FROM THE CARIACO TRENCH SEDIMENTS

Graeme L. Lyon

Institute of Nuclear Sciences
Department of Scientific and Industrial Research
Lower Hutt, New Zealand

ABSTRACT

The 198 meter core of Quaternary sediments obtained from Hole 147, Leg 15 of the Deep Sea Drilling Project was laid down under anoxic conditions in the Cariaco Trench below 882 meters of water off the coast of Venezuela. Pockets of gas from core depths between 45 and 180 meters were found to be 86-99% methane. The carbon dioxide content of the ten pockets sampled reached a maximum at about 130 meters.

Isotopic analysis of the methane showed $\delta^{13}C_{PDB}$ values between -59 and -76°/oo. This is consistent with a biological origin due to the anaerobic reduction of organic matter in the sediments. The D/H ratios of the methane in all samples were similar, with a mean δD_{SMOW} value of -178°/oo. In carbon dioxide $\delta^{13}C_{PDB}$ and $\delta^{18}O_{PDB}$ both fell in the range of +0.5°/oo to -4.5°/oo, and both values approached zero on the PDB scale as depth increased.

It might be expected that the methane would have had time to reach isotopic equilibrium with the carbon dioxide and interstitial water. The isotopic fractionations between CH_4 and CO_2 for $^{13}C/^{12}C$ and between CH_4 and H_2O for D/H were compared with equilibrium fractionations calculated from thermodynamic data. The trench temperature of 17°C differs from thermodynamic equilibrium temperatures by about 40°C.

INTRODUCTION

Site 147 of Leg 15 of the Deep Sea Drilling Project was at

91

10°42.65'N, 65°10.46'W in the Cariaco Trench off the coast of
Venezuela. This trench is an elongated depression surrounded by a
sill at 150 meters which restricts exchange of water with the deep
ocean. This restriction has produced anaerobic isothermal (16.9°C)
bottom water conditions [*Richards, 1960*]. Holes were drilled in
882 meters water depth at the site, and a total of 198 meters of
core was taken. At core depths below 40 meters, gas pockets were
found and sampled [*Horowitz et al. 1973*]. These gases were mostly
methane [*Hammond et al. 1973*], and aliquots were sent to this laboratory for isotopic analyses.

Analytical methods have been described elsewhere [*Lyon, 1973*].
Stable isotope results measured relative to laboratory standards
have been converted to the international standards: PDB for
$^{13}C/^{12}C$ and $^{18}O/^{16}O$ [*Craig, 1957*] and SMOW for D/H [*Craig, 1961*].
Values are quoted as per mille ($^\circ/_{oo}$) difference from the standards,
with standard deviations of $\pm0.3^\circ/_{oo}$ for $\delta^{13}C$ and $\delta^{18}O$, and
$\pm3^\circ/_{oo}$ for δD.

RESULTS AND DISCUSSION

The data obtained from the analyses are presented in Table 1.

TABLE 1.   Isotopic Analyses of Gas, DSDP Leg 15

JOIDES Sample no.	Depth in Sediment (m)	% CO_2	$\delta^{18}O_{PDB}$ CO_2	$\delta^{13}C_{PDB}$ CO_2	CH_4	δD_{SMOW} CH_4
147-6-4	47	1.0	n.d.*	n.d.	-76.3	n.d.
147B-6-2	51	2.4	n.d.	-2.9	-67.6	-179
147B-7-4	63	2.7	n.d.	-4.2	-65.7	-173
147-10	82	3.0	-2.9	-2.9	-64.6	-175
147B-9-4	84	3.7	-2.2	-2.2	-60.8	n.d.
147B-11-3	100	12.4	-0.3	0.0	-65.8	-179
147-14-3	120	4.7	-3.4	-1.5	-59.6	-179
147C-2-2	128	13.2	+0.4	+0.2	-65.1	-181
147C-4-4	148	11.1	-0.3	+0.5	-64.1	-183
147C-7-3	175	7.5	-0.4	-0.5	-64.6	-176

*n.d. = not determined

The carbon dioxide fraction (as a percentage of $CO_2 + CH_4$) is seen to generally increase with depth in the sequence, as also found by *Hammond et al.* [*1973*]. Some carbon dioxide samples were too small for isotopic analysis, but in general, both the $\delta^{13}C$ and $\delta^{18}O$ values for the carbon dioxide are nearer to zero (on the PDB scale) as depth increases. However, non-equilibrium outgassing of the interstitial water could account for some of the variation in these isotopic ratios. For oxygen isotopic equilibrium with water with the composition of Standard Mean Ocean Water (SMOW), the $\delta^{18}O$ values would indicate temperatures between 24 and 47°C [*O'Neil and Adami, 1969*], but these may have been affected by equilibration with water during and subsequent to extraction.

The methane isotopic data show carbon that is extremely depleted in ^{13}C, which is typical for the products of anaerobic microorganisms [*Claypool and Kaplan, 1974*]. The measured values of $\delta^{13}C$ are in the range -59 to -76°/$\circ\circ$, and they are similar to the measured values found by *Claypool et al.* [*1973*] in the same sediments. The deuterium contents of the methane are all essentially identical within experimental error, with a mean δD value of -178°/$\circ\circ$ relative to SMOW. This value is within the range -165 to -219°/$\circ\circ$, which *Schiegl and Vogel* [*1970*] have found for methane in natural gas from Germany and from South African coal measures. However, those gases had $\delta^{13}C$ values of approximately -40°/$\circ\circ$ (J.C. Vogel, personal communication).

Gunter and Musgrave [*1971*] found lower deuterium concentrations in geothermal methane collected from Yellowstone National Park, U.S.A., and these data are shown in Figure 1. Also shown on the graph is one New Zealand natural gas, within the range shown by *Schiegl and Vogel* [*1970*] and J.C. Vogel (personal communication). The New Zealand and Kenyan geothermal gas data are unpublished values from the Institute of Nuclear Sciences, New Zealand. The New Zealand methane sample, marked 49°C, is from a warm spring in a swamp near Maketu, and gave a measured $\delta^{13}C$ value of -71°/$\circ\circ$. This gas also had no detectable hydrogen and very little carbon dioxide; thus it is probably biogenic rather than truly geothermal. The Deep Sea Drilling samples from Hole 147 are seen to form an entirely separate group. Deuterium concentration differences in methane from different sources may be related to the organic matter and to D/H ratios in the water from marine and nonmarine environments, but little data are available. *Schiegl and Vogel* [*1970*] recorded land plants with δD values between -30 and -120°/$\circ\circ$, and for marine organisms δD ranged between -10 and -170°/$\circ\circ$. They were able to relate the deuterium content of local precipitation with the δD for wood. In general, the deuterium content of geothermal methane also appears to be related to precipitation.

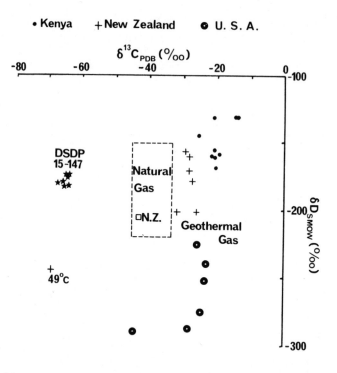

Fig. 1. $\delta^{13}C$ - δD *Plot for Naturally Occurring Methane.* The rectangle outlined with broken lines encloses $\delta^{13}C$ and δD values measured by *Schiegl and Vogel* [*1970*] on natural gas from Germany and South Africa.

Approximate values of δD for local precipitation are: Yellowstone -140°/₀₀ [*Craig, 1963*], New Zealand -40°/₀₀ (this laboratory) and Kenya about 0°/₀₀.

Recently, *Nissenbaum et al.* [*1972*] suggested that methane in sediments is due to reduction of carbon dioxide by bacteria using molecular or organically available hydrogen. *Presley and Kaplan* [*1968*] had shown that the total dissolved carbon dioxide (CO_3^{2-} + HCO_3^- + H_2CO_3 + CO_{2aq}) initially forms from organic matter in the sediments with the same $\delta^{13}C$ value as the organic matter (about -20°/₀₀). Subsequent reduction to methane with a ^{13}C fractionation factor of up to 1.08 increases the ^{13}C concentration in the remaining total dissolved carbon dioxide. The actual fractionation would depend on kinetic effects, but $\delta^{13}C$ values measured in this study (about zero for carbon dioxide and -70°/₀₀ for methane) are consistent with the suggestion of *Nissenbaum et al.* [*1972*]. These authors showed that values of $\delta^{13}C$ for the total dissolved carbon dioxide in Saanich Inlet sediments could vary from $-11°/oo$ to $+17°/oo$, though carbon dioxide gas may have been lost from their samples during collection and storage.

Bottinga [1969] has calculated equilibrium fractionation factors from carbon and hydrogen isotopes between various molecular species, including carbon dioxide, methane and water vapor. These could exchange by the reversible reaction $CO_2 + 4H_2 \rightleftharpoons CH_4 + 2H_2O$. For the present samples, the ^{13}C fractionation between carbon dioxide and methane indicate equilibrium temperatures between 27 and 52°C. For the samples collected below 100 meters in the sediments, the methand and carbon dioxide $\delta^{13}C$ values are relatively more constant, indicating temperatures of 27 to 34°C. These values do not differ from the core temperature (17°C) by more than 35°C, whereas geothermal gases [*Hulston and McCabe, 1962; Lyon and Hulston, 1971*] typically show Bottinga's calculated temperatures to be 140°C above the maximum measured bore-hole temperatures. It may be that the equilibrium reaction, which is known to be very slow, has only allowed equilibrium to be reached in the oldest sediments of this sequence. But $\delta^{13}C$ data show that dissolved carbon dioxide and calcium carbonate in the sediment have not equilibrated [*Claypool et al., 1973*].

The deuterium concentration in the methane could equilibrate with the interstitial water, which has a composition close to SMOW $(+5$ to $+11°/oo)$ [*Friedman and Hardcastle, 1973*]. Using data of J.R. Hulston (personal communication), who has revised the calculations of *Bottinga* [1969] and *Merlivat and Nief* [1967], δD values of $-178°/oo$ for methane and of $+8°/oo$ for the interstitial water, calculate to an equilibrium temperature of $-23°C$, which differs by about 40°C from the estimated true temperature within the core. Subsequent refinements in the calculations or experimentally determined fractions may allow such measurements to be used as a thermometer in the future.

Chemical equilibrium in the system $CH_4-H_2-CO_2-H_2O$ should also be attained, and if partial pressures of methane, hydrogen, carbon dioxide, and the total pressure in the sediment were measured, the calculations of *Ellis* [1957] could be used to obtain chemical equilibrium temperatures in the sediments. These calculations show that for equilibrium at high temperature (above 600°C) methane concentrations are negligible, but for temperatures below 100°C, hydrogen concentrations would be very low. The lack of detectable hydrogen [*Hammond et al. 1973*] is therefore consistent with thermodynamic equilibrium, and improved methods of collection and measurement [*Hammond et al. 1973*] may produce quantitative values. The possibility of being able to distinguish between isotopic equilibrium fractionation, and the kinetic fractionation observed in the very much younger Saanich Inlet sediments [*Nissenbaum et al. 1972*] should be pursued, and estimates of the rate of equilibration under natural, low temperature conditions may eventually be determined.

ACKNOWLEDGMENTS

The author wishes to thank Mrs. M.A. Cox for her invaluable technical assistance, and Dr. J.R. Hulston for useful discussions.

Institute of Nuclear Sciences Contribution No. 578.

REFERENCES

Bottinga, Y., Calculated fractionation factors for carbon and hydrogen isotope exchange in the system calcite-carbon dioxide-graphite-methane-hydrogen-water vapor, *Geochim. et Cosmochim. Acta, 33, 49*, 1969.

Claypool, G. E., and I. R. Kaplan, The origin and distribution of methane in marine sediments, in *Natural Gases in Marine Sediments*, edited by I. R. Kaplan, pp. 99-139, Plenum Press, New York, 1974.

Claypool, G. E., I. R. Kaplan, and B. J. Presley, *Initial Reports of the Deep Sea Drilling Project*, vol. 19, pp. 879-884, U. S. Government Printing Office, Washington, 1973.

Craig, H., Isotopic standards for carbon and oxygen and correction factors for mass spectrometry analysis of carbon dioxide, *Geochim. et Cosmochim. Acta, 12, 133*, 1957.

Craig, H., Standard for reporting concentrations of deuterium and oxygen-18 in natural waters, *Science, 133, 1833*, 1961.

Craig, H., The isotopic geochemistry of water and carbon in geothermal areas, in *Nuclear Geology on Geothermal Areas*, edited by E. Tongiorgi, p. 17, Consiglio Nazionale delle Ricerche Laboratorio di Geologia Nucleare, Pisa, 1963.

Ellis, A. J., Chemical equilibrium in magmatic gases, *Amer. J. of Sci., 255, 416*, 1957.

Friedman, I., and K. Hardcastle, *Initial Reports of the Deep Sea Drilling Project*, vol. 20, pp. 901-904, U. S. Government Printing Office, Washington, 1973.

Gunter, B. D., and B. C. Musgrave, New evidence on the origin of methane in hydrothermal gases, *Geochim. et Cosmochim. Acta, 35, 113*, 1971.

Hammond, D. E., R. Horowitz, W. S. Broecker, and R. Bopp, *Initial Reports of the Deep Sea Drilling Project*, vol. 20, pp. 765-772, U. S. Government Printing Office, Washington, 1973.

Horowitz, R. M., L. S. Waterman, and W. S. Broecker, *Initial Reports of the Deep Sea Drilling Project*, vol. 20, pp. 757-764, U. S. Government Printing Office, Washington, 1973.

Hulston, J. R., and W. J. McCabe, Mass spectrometer measurements in the thermal areas of New Zealand. Part 2. Carbon isotopic ratios, *Geochim. et Cosmochim. Acta, 26,* 399, 1962.

Lyon, G. L., *Initial Reports of the Deep Sea Drilling Project,* vol. 20, pp. 773-774, U. S. Government Printing Office, Washington, 1973.

Lyon, G. L., and J. R. Hulston, The application of stable isotope measurements to geothermal temperature studies, paper presented at the 43rd Congress, Australian and New Zealand Association for the Advancement of Science, Brisbane, May 1971.

Merlivat, L., and G. Neif, Fractionnement isotopique lors des changements d'état solide-vapeur et liquid-vapeur de l'eau à des températures inférieures à 0°C, *Tellus, 19,* 122, 1967.

Nissenbaum, A., B. J. Presley, and I. R. Kaplan, Early diagenesis in a reducing fjord, Saanich Inlet, British Columbia, Part 1. Chemical and isotopic changes in major components of interstitial water, *Geochim. et Cosmochim. Acta, 36,* 1007, 1972.

O'Neil, J. R., and L. H. Adami, The oxygen isotope partition function ratio of water and the structure of liquid water, *J. Phys. Chem., 73,* 1553, 1969.

Presley, B. J., and I. R. Kaplan, Changes in dissolved sulfate, calcium and carbonate from interstitial water of near-shore sediments, *Geochim. et Cosmochim. Acta, 32,* 1037, 1968.

Richards, F. A., Some chemical and hydrographic observations along the north coast of South America, Part 1. Cabo Tres Puntas to Curacao, including the Cariaco Trench and the Gulf of Cariaco, *Deep-Sea Res., 7,* 163, 1960.

Schiegl, W. E., and J. C. Vogel, Deuterium content of organic matter, *Earth Planet. Sci. Lett., 7,* 307, 1970.

THE ORIGIN AND DISTRIBUTION OF METHANE IN MARINE SEDIMENTS

George E. Claypool* and I.R. Kaplan

Institute of Geophysics and Planetary Physics
University of California at Los Angeles
Los Angeles, California 90024

ABSTRACT

Methane has been detected in several cores of rapidly deposited (> 50 m/my) deep sea sediments. Other gases, such as carbon dioxide and ethane, are commonly present but only in minor and trace amounts, respectively. The methane originates predominantly from bacterial reduction of CO_2, as indicated by complimentary changes with depth in the amount and isotopic composition of redox-linked pore water constituents: sulfate-bicarbonate and bicarbonate-methane.

Presently, no precise determination exists of the amount of gas present under *in situ* conditions in deep sea sediments. Using C^{13}/C^{12} isotope ratios of the dissolved bicarbonate and methane, and employing kinetic calculations based on Rayleigh distillation equations, the amounts of methane generated by reduction of carbon dioxide by hydrogen has been estimated. The amounts calculated suggest that a minimum of 20 mmol CH_4/kg interstitial water is formed.

A methane concentration of 20 mmol/kg approaches the amount required for the formation of gas hydrates under pressure-temperature conditions corresponding to a water column of about 500 meters, with a temperature of $5°C$ at the sediment-water interface. Depth of stability of the gas hydrate within the sediment is directly proportional to: hydrostatic pressure, or height of the water column above the sediment, temperature at the sediment surface, the geothermal gradient, and concentration of methane. Under average oceanic conditions, gas hydrates could be stable in sediment under a 3 km water column to depths of approximately 600 meters, if sufficient methane is present.

---

*Present address Branch of Oil and Gas Resources, U.S. Geological Survey, Denver, Colorado 80225.

Gas hydrates have been proposed as the cause of anomalously high acoustic velocities in the upper 500-600 meters of sediment in the Blake-Bahama outer ridge. It is here suggested that acoustic reflectors in gas-rich sediment is associated with temperature-dependent lithologic transitions, which are in part formed by diagenetic processes involving microbiological methane generation.

Under certain conditions, carbonate ion must be removed from solution during methane production to maintain pH equilibrium between the pore water and the sediment. Authigenic carbonates, typically iron-rich nodules and cements, have been observed in the zone of active methane production. This link between methane production and carbonate precipitation may be an important mechanism for lithification of deep sea sediments.

INTRODUCTION

Methane originates from the decomposition of organic matter in the absence of oxygen. At high temperatures (> 50°C), methane and other hydrocarbons are produced by non-biological, thermocatalytic reactions. In recent sediments, methane is one end product of the anaerobic bacterial decomposition of organic matter.

Physiological and ecological constraints limit the extent of bacterial methane production under present-day earth surface conditions. Methane-producing bacteria are strict anaerobes and will not grow in the presence of even traces of oxygen. In addition, there is some evidence that methane-producing bacteria do not ordinarily grow in the presence of dissolved sulfate. These limitations effectively confine the activity of these microorganisms to highly specialized and restricted environments, such as: dung heaps and anaerobic sewage digestors [*Toerien and Hattingh, 1969*]; the digestive tract of animals, especially ruminants [*Bryant, 1965*], but also including man [*Steggerda and Dimmick, 1966*]; poorly-drained swamps, bays, paddy fields and anoxic fresh-water lake bottoms [*Oana and Deevey, 1960; Koyama, 1964*]; and marine sediments beneath the zone of active sulfate reduction [*Emery and Hoggan, 1958*].

Methane production is easily detected in bogs, swamps, and brackish-water marshes because it occurs close to the surface, and may be easily detected as it issues from the sediment. In marine sediments, sulfate reduction appears to be the dominant observable process of anaerobic bacterial respiration at shallow burial depths [*Zobell and Rittenberg, 1948*]. Evidence for bacterial methane production in marine sediments is usually not detected at shallow depths unless dissolved sulfate is almost completely removed from the interstitial water [*Nissenbaum et al. 1972*]. In deep-sea sediments, complete removal of dissolved sulfate does not typically occur until burial depths of the order of some tens of meters have been achieved

[*Sayles et al. 1970, 1972, 1973a,b; Presley and Kaplan, 1970; Manheim et al. 1970, 1973; Sayles and Manheim, 1971; Waterman et al. 1973*]. Consequently, it has been difficult to observe the geochemical effects of methane production in open marine sediments sampled by conventional coring methods. With the advent of deep coring by the Deep Sea Drilling Project (DSDP), information on the distribution of methane has become available [*Claypool et al. 1973*].

In this report, we briefly review the biological process of methane formation and some published descriptions of sedimentary methane occurrence. We also report on the chemical and isotopic composition of gas samples from DSDP cores, and discuss some of the diagenetic effects of methane generation and occurrence in marine sediments.

MICROBIAL METABOLISM IN MARINE SEDIMENTS

Sedimentary environments in which organic matter is deposited at a rate exceeding the supply of dissolved oxygen, are characterized by the appearance at some depth of anaerobic (or more precisely, anoxic) conditions. This depth, below which there is no dissolved oxygen, marks the boundary between regimes of aerobic and anaerobic metabolism. The aerobic-anaerobic boundary may be above, but more commonly is below or at, the sediment-water interface.

Aerobic respiration using organic matter is the most efficient energy-yielding metabolic process. In the absence of molecular oxygen, other oxidized substances are used as terminal acceptors for electrons generated in the degradation of organic matter. Table 1 summarizes some common energy-yield metabolic processes and the approximate maximum amount of energy that can be derived from them. The ΔG^0 values are for standard conditions (1 molar concentration) and are included for comparative purposes. Actual free energy changes under natural conditions may be different.

Two general types of metabolic processes are shown in Table 1; one type utilizes inorganic compounds (i.e., O_2, NO_3^-, $SO_4^=$, HCO_3^-) as electron acceptors, and the other type uses organic compounds (usually intermediates produced in the dissimilation of the substrate itself as electron acceptors). The former are considered to be respiratory processes (both aerobic and anaerobic) and the latter fermentative processes. In the marine environment, both types of metabolism are carried out primarily by bacteria, organisms that are ubiquitous in the biosphere. The local and temporal environment selects for the specific composition of the bacterial population. The relative energy yields listed in Table 1 are a theoretical measure of the selective advantage conferred on the organisms capable of catalyzing the various processes. When two or more physiologically-specialized types of microorganisms compete for the same

TABLE 1   Energy-yielding Metabolic Processes as Coupled
Oxidation-reduction Reactions

	$\Delta G^{\circ\prime}$ (kcal per mole of glucose equivalent oxidized)
a. Aerobic respiration	
$CH_2O + H_2O \rightarrow CO_2 + 2H_2$	
$2H_2 + O_2 \rightarrow 2H_2O$	
$\overline{CH_2O + O_2 \rightarrow CO_2 + H_2O}$	-686
b. Nitrate reduction	
$5CH_2O + 5H_2O \rightarrow 5CO_2 + 10H_2$	
$10H_2 + 4NO_3^- + 4H^+ \rightarrow 2N_2 + 12H_2O$	
$\overline{5CH_2O + 4NO_3^- + 4H^+ \rightarrow 2N_2 + 5CO_2 + 7H_2O}$	-579
c. Sulfate reduction	
$2CH_2O + 2H_2O \rightarrow 2CO_2 + 4H_2$	
$4H_2 + SO_4^= \rightarrow S^= + 4H_2O$	
$\overline{2CH_2O + SO_4^= \rightarrow S^= + 2CO_2 + 2H_2O}$	-220
d. Carbonate reduction	
$2CH_2O + 2H_2O \rightarrow 2CO_2 + 4H_2$	
$4H_2 + HCO_3^- + H^+ \rightarrow CH_4 + 3H_2O$	
$CO_2 + H_2O \rightarrow HCO_3^- + H^+$	
$\overline{2CH_2O \rightarrow CH_4 + CO_2}$	-99
e. Nitrogen fixation	
$3CH_2O + 3H_2O \rightarrow 3CO_2 + 6H_2$	
$6H_2 + 2N_2 \rightarrow 4NH_3$	
$\overline{3CH_2O + 2N_2 + 3H_2O \rightarrow 4NH_3 + 3CO_2}$	-57
f. Fermentation: heterolactic	
glucose $\rightarrow$ acetaldehyde $+ CO_2 +$ lactate $+ H_2$	
$H_2 +$ acetaldehyde $\rightarrow$ ethanol	
$\overline{\text{glucose} \rightarrow \text{lactate} + \text{ethanol} + CO_2}$	-49
g. Fermentation: Stickland reaction	
alanine $+ 2H_2O \rightarrow NH_3 +$ acetate $+ CO_2 + 2H_2$	
$2H_2 + 2$ glycine $\rightarrow 2NH_3 + 2$ acetate	
$\overline{\text{alanine} + 2 \text{ glycine} + 2H_2O \rightarrow 3 \text{ acetate} + CO_2 + 3NH_3}$	-17

ecological niche (i.e., for the same organic substrate supply), organisms capable of deriving the greatest metabolic energy will dominate. As a result, a succession of organisms should be detected in time

(equivalent to depth down a sediment column), with the oxygen utilizing organisms at the sediment surface and the less efficient
species at greater depths.

Although glycolytic and amino acid fermentations yield relatively little energy compared to aerobic and anaerobic respirations,
they are not competitive at the same level of substrate organization
and can occur simultaneously. In fact, fermentation processes are
preparatory steps for the respiration processes, degrading organic
monomers such as sugars and amino acids to smaller (two-, three-
and four-carbon) organic acids and alcohols. It is at this lower
level of substrate organization where the potential exists for competition among ecosystems organized around different types of respiratory processes. Other inhibiting factors, such as compounds
that are toxic to one genus and not to another, may exert a more
direct selective control; but even in the absence of such positive
inhibitors, the competitive exclusion that arises through more efficient substrate utilization apparently results in mutually-ex-
clusive metabolic processes [see *Mechalas, 1974* for further discus-
sion].

Ecological Succession

The result is the succession of microbial ecosystems, illustrated
in Figure 1, showing a generalized cross section of an open marine,
organic-rich sedimentary environment. The interrelations between
sedimentological and ecological factors bring about three distinct
biogeochemical environments, each characterized by the dominant
form of respiratory metabolism. These three zones are: the aerobic
zone; the anaerobic sulfate-reducing zone; and, the anaerobic methane-producing (or carbonate-reducing) zone. The presence of zones
or layers of the sediment characterized by successively less efficient modes of respiratory metabolism can be inferred on the basis
of changes in concentration and isotopic composition of the respiratory metabolites, discussed below. In each of these zones, the
dominant microbial population exploit the environment, creating a
new environment that favors other species. Thus, the transitions
from aerobic sediment, to anaerobic sulfate-reducing sediment, to
anaerobic methane-producing (or carbonate-reducing) sediment, are
geochemical consequences of species-induced environmental changes,
i.e., the depletion and exhaustion of the required respiratory me-
tabolite and the generation of new products [*Hardin, 1966*].

When such an ecological sequence as that depicted in Figure 1
is established, the biogeochemical zones move upward with time,
keeping pace with the rate of addition of sediment at the water-
sediment interface, or alternatively, the sediment moves downward
through a succession of diagenetic environments. When oxygen has
been depleted, obligatory aerobic organisms cease to grow. Some

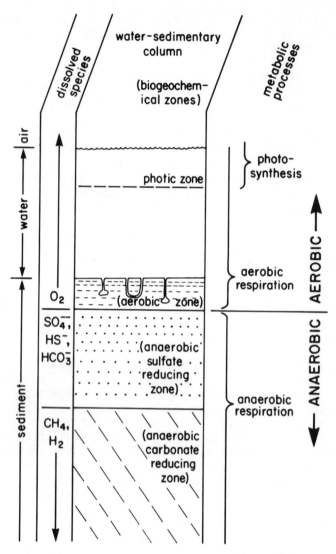

Fig. 1. An idealized cross section of an open marine organic-rich (reducing) sedimentary environment. The biogeochemical zones are a consequence of ecological succession.

microorganisms, the facultative anaerobes, can switch from aerobic respiration to fermentation or anaerobic respiration using nitrate

[*Vaccaro, 1965*], or inorganic sulfur compounds other than sufate [*Tuttle and Jannasch, 1973*] as alternative energy-yielding metabolic processes. However, as these electron acceptors are limited, this zone is narrow and very restricted. Sulfate-reducing bacteria become dominant with the onset of anaerobic conditions in marine sediments, primarily because of the relatively high concentration of sulfate (0.028 molar) in normal sea water. In addition, it is probable that only a few other microbial species can tolerate high contents of H_2S, which is an end product. Sulfate-reducing bacteria are limited in their range of oxidizable substrates, utilizing chiefly lactic acid and four-carbon dicarboxylic acids, and in some species and strains, molecular hydrogen [*Postgate, 1965*]. The abundance of these compounds in the sediment at any given time is probably small; thus an active population of sulfate-reducing bacteria requires a symbiotic association with an anaerobic fermenting population in order to provide a source of oxidizable carbon substrate.

In situ bacterial sulfate reduction can be followed by depletion of dissolved sulfate in interstitial water and relative enrichment in S^{34} (δS^{34} becoming more positive) as a result of preferential S^{32} removal in the sulfide. An example of this process is illustrated in Figure 2, from data obtained on sediments of hole 148 DSDP [*Presley et al. 1973*].

Bacterial Methane Production

When the sulfate concentration of the water buried with the sediment is low, as in brackish to fresh water environments, or in marine sediments below the zone of sulfate reduction, carbonate or CO_2 reduction should replace sulfate reduction as the preferred process of anaerobic respiration. Whether these two processes overlap, or are mutually exclusive, is presently unknown. However, high contents of methane have not been reported in sediments containing dissolved sulfate, and methane production appears to begin immediately upon the disappearance of sulfate. One possible explanation is that H_2S is toxic to methane bacteria, although inhibition of this type is not consistent with some observations of apparent methane production in the presence of unreacted HS^-. An alternative explanation is that free hydrogen is not available for CO_2 reduction in the presence of dissolved sulfate. Many sulfate-reducing bacteria have a hydrogenase electron transport system and can utilize, but do not require, free hydrogen. In contrast, methane bacteria depend on interspecies electron transfer in the form of dissolved hydrogen [*Bryant et al. 1967; Iannotti et al. 1973*] as their energy source. It may be that an adequate partial pressure of hydrogen gas cannot be generated in the presence of sulfate ion during active sulfate reduction.

In addition to these environmental restrictions, the methane

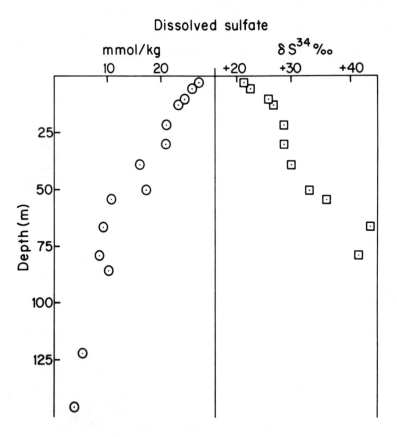

Fig. 2.   Changes with depth in the concentration and sulfur isotope ratio (δS^{34}) of dissolved sulfate in the interstitial water of Aves Ridge (DSDP site 148) sediments, caused by the process of bacterial sulfate reduction [*Presley et al. 1973*, and unpublished data].

producing bacteria are also quite circumscribed with respect to utilizable substrates for methane formation and growth. Hydrogen and carbon dioxide are the preferred substrates for eight species of methanogenic bacteria that have been studied in pure culture [*Wolfe, 1971*]. Formate can also be used by most of these, but is probably first converted to hydrogen and carbon dioxide. One species (*Methanosarcina barkerii*) can use methanol or acetate in addition to hydrogen and carbon dioxide, but growth is much slower, especially with acetate [*Wolfe, 1971*].

Evidence based on growth experiments using $C*H_3COOH$ labeled substrate and raw sewage, favors the direct formation of methane and CO_2 from acetic acid [*Smith and Mah, 1966*]. However, the responsible organism(s) that can carry out this one-step conversion has/have not yet been isolated, and it is conceivable that some intermediate step is involved. Acetic acid and some other low molecular weight carboxylic acids have been analyzed in marine sediment [see Table 2; *Hoering, 1968*], but their concentration is very low, and conversion to methane would only permit formation of a few mmol CH_4. Furthermore, acetic acid is thought to form as an end product of sulfate reduction, yet methane appears to form only after sulfate depletion. This suggests that in marine sediments acetate is utilized in metabolic systems other than methane production. Therefore, whereas reduction of CO_2 *per se* may not be the only mechanism of biological methane production, it is energetically the most favorable when hydrogen is available, and probably accounts for most of the methane produced in marine environments.

Because of physiological limitations, methane producing bacteria are incapable of direct metabolism of either the usual carbohydrate substrates, such as cellulose and monosaccharides, or fatty acids larger than acetate. Rather, a mixed population exists that includes organisms capable of breaking down the more abundant complex organic molecules into the simple compounds required by the methane producers. Hydrogen producing bacteria are critical members of the anaerobic ecosystem, but there is even less information concerning these species than for the methane producers. It seems likely that there are two types of hydrogen producers:

1) Specialized organisms that catalyze energetically unfavorable oxidations, such as

$$CH_3CH_2OH + H_2O \rightarrow CH_3COOH + 2H_2$$

when associated with symbiotic methane producers [*Bryant et al. 1967*];

2) A broad range of fermenting bacteria that can produce hydrogen, instead of a more reduced organic compound, if hydrogen-utilizing species are also present [*Iannotti et al. 1973*].

In both cases, organic degradation is accelerated by removal of hydrogen, which in turn depends on the availability of suitable electron acceptors, such as CO_2.

TABLE 2. Volatile Aliphatic Acids Isolated from Sediments of Offshore Southern California Basins

Acid	San Nicholas Basin		San Pedro Basin		Tanner Basin	
	ppm*	mmol/kg[†]	ppm*	mmol/kg[†]	ppm*	mmol/kg[†]
acetic (C_2)	1810	2.1	1350	1.6	459	0.5
propanoic (C_3)	97	0.09	179	0.17	37	0.03
i-butanoic (C_4)	13	0.01	36	0.03	13	0.01
n-butanoic (C_4)	4		40		11	
i-pentanoic (C_5)	8		46		16	
n-hexanoic (C_6)	32		63		24	
n-heptanoic. (C_7)	41		113		13	

*Relative to organic matter in the sediment.

[†]Relative to the pore water of the sediment, assuming 7% organic matter and 50% water. Adopted from *Hoering [1968]*.

TABLE 3. Possible Mechanisms of Methane Carbon
Isotope Fractionation

	Reference
I. Equilibrium	*Craig* [1953] *Bottinga* [1969]
II. Kinetic	
A. Migration/diffusion	*Columbo et al.* [1966]
B. C^{12}-C^{12} versus C^{12}-C^{13} bond-breaking in thermal cracking	*Sackett et al.* [1968]
C. Biogenic C^{12} enrichment	*Rosenfeld and Silverman* [1959]

CARBON ISOTOPE FRACTIONATION AND METHANE GENERATION IN MARINE SEDIMENTS

The mechanisms responsible for carbon isotope fractionation are outlined in Table 3. Two processes can affect the δC^{13} of methane: equilibrium fractionation between methane and other carbon compounds; and kinetic effects associated with methane formation by thermal cracking or through bacterial action, and isotopic separation associated with gas migration. Of the possible mechanisms, only the kinetic effects of microbial metabolism are considered in the interpretation of the δC^{13} of methane produced during the early diagenetic history of sediments. Other kinetic and equilibrium mechanisms will not be considered because the conditions required to produce significant C^{12} enrichment in methane are not fulfilled on the basis of the known history of the sediments under consideration.

Here, we intend to examine the environments of methane production (i.e., water depth and depth of burial), point out the observed amounts and δC^{13} relations for both methane and dissolved carbonate, and show how δC^{13} measurements can be used to infer the extent of biological methane production in the absence of reliable concentration measurements.

Dissolved Methane and Interstitial Water Chemistry

Some environmental conditions under which methane has been detected are summarized in Table 4, for coastal marshes of Southern Louisiana [see *Whelan, 1974*], Chesapeake Bay [*Reeburgh, 1969*], and Southern California marshes and offshore basins [*Emery and Hoggan, 1958*]. In coastal marshes, the partial pressure of dissolved methane must exceed a total pressure of just one atmosphere before bubble formation is possible. This corresponds to a dissolved methane concentration of about 30 ml/l (STP) of interstitial water. In the shallow (60 cm) cores of the northernmost marsh sediment analyzed by Whelan, the methane concentration approached this limit where the chlorinity, and presumably the sulfate concentration, was low. Where sulfate may have been present, as suggested by high chlorinities or detectable H_2S, methane is present in much lower amounts, as shown by the analysis of gas from sediment of the southernmost South Louisiana coastal marsh by *Whelan* [*1974*] and the Newport Bay and Seal Beach marshes analyzed by *Emery and Hoggan*, [*1958*].

Reeburgh [*1969*] measured methane in Chesapeake Bay sediment at water depths of 15.2 meters and 30 meters. At both locations, the maximum methane concentration measured is about the same as the theoretical solubility limit. Reeburgh has presented additional evidence that methane concentration in these sediments is controlled by its solubility in the pore water. Sediment with the highest methane contents has lost nitrogen and argon by gas stripping associated with methane ebullition. The highest concentrations of methane quantitatively measured in sediment pore waters were those observed by *Emery and Hoggan* [*1958*] in Santa Barbara Basin. The observed maximum methane concentration (10 mmol/kg) in Santa Barbara Basin sediment under about 585 meters of water, is also close to the theoretical saturation limit, because these sediments are within the P-T region of gas hydrate stability. Therefore, methane has probably not escaped from these sediments by bubbling out, at least from the depth intervals sampled.

In Figure 3, the methane concentrations measured by Emery and Hoggan and the sulfate concentrations reported by *Sholkovitz* [*1973*] for another core from the central deep of Santa Barbara Basin are plotted against depth. If the trend of sulfate decrease is extrapolated to zero, it intersects the trend of methane increase at a very low concentration, which is consistent with the suggestion that sulfate reduction and methane production are, to a large degree, mutually exclusive.

Total Dissolved CO_2 (ΣCO_2)

The concentration and the δC^{13} of pore water bicarbonate,

TABLE 4. Environments of Bacterial Methane Production

Water Depth	Chlorinity	Temperature	Burial Depth	Methane Content	
				Observed Maximum	Theoretical Solubility
(m)	(°/oo)	(°C)	(m)	(mmol/kg)	(mmol/kg)
Southern Louisiana Coastal Marsh [*Whelan 1974*]					
≈ 0.3	6	20	0.5	1.1	1.3
≈ 0.3	12	20	0.5	0.01	1.2
Southern California Coastal Marsh [*Emery and Hoggan 1958*]					
≈ 0.3	20	20	0.2	0.01	1.1
≈ 0.3	21	20	0.3	0.002	1.1
Chesapeake Bay Sediments [*Reeburgh 1969*]					
15.2	10	15	1.0	4.3	4.3
30	10	15	0.9	6.5	6.5
Santa Barbara Basin [*Emery and Hoggan 1958*]					
585	19	5	3.8	10.4	10 to 20*

*Approximate range for solution in equilibrium with methane hydrate [*Makogon et al. 1972*].

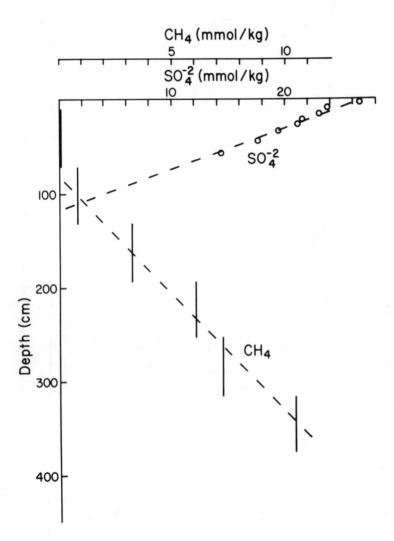

Fig. 3. Changes with depth in the concentration of dissolved sulfate [*Sholkovitz, 1973*] and methane [*Emery and Hoggan, 1958*] in sediments of Santa Barbara Basin. The extrapolated sulfate trend is consistent with mutually exclusive processes of microbial anaerobic respiration.

measured as ΣCO_2, are sensitive indicators of the degradation of organic matter. During sulfate reduction, isotopically light $(-20°/_{oo})$ bicarbonate is added to the pore water, as shown by *Presley and Kaplan* [*1968*], whose data for a core in Santa Catalina Basin are reproduced in Figure 4. In this core, sulfate does not go to zero over the depth interval sampled; thus methane production would not be anticipated. The isotopically light bicarbonate in the presence of sediment carbonate with $\delta C^{13} \simeq 0$ suggest that no exchange in carbon is occurring.

The effect of methane production on the δC^{13} of dissolved bicarbonate was demonstrated by *Nissenbaum et al.* [*1972*] in the sediments of Saanich Inlet, an anoxic fjord on the coast of British Columbia. Their data for a typical core are reproduced in Figure 5. In these sediments, sulfate reduction is so rapid that methane production begins at very shallow burial depths. Here, during methane formation, $HC^{12}O_3^-$ is removed at a rate about 7% faster than the $HC^{13}O_3^-$ removal rate [*Rosenfeld and Silverman, 1959*]. The result is a marked C^{13} enrichment in the residual interstitial water bicarbonate. In Saanich Inlet sediments, the rate of anaerobic oxidation or organic matter and the addition of $-20°/_{oo}$ CO_2 is, however, faster than the rate of CO_2 removal in methane production; hence the bicarbonate concentration continues to increase.

A more typical example of the effects of both sulfate reduction and methane production on the interstitial water chemistry is shown in South Guymas Basin, Gulf of California. Pore water analyses obtained from *M.B. Goldhaber* [*1974*] are plotted in Figure 6. In these sediments, sulfate concentration goes to zero at about 2 meters depth. At this point, the onset of methane production is reflected in decreased titration alkalinity and increased δC^{13} of total dissolved CO_2. Direct evidence of methane production was provided by a gas pocket formed in the core liner at above 2.2 meters when the cored sediments were brought to the surface. This gas pocket contained methane with a δC^{13} of $-76°/_{oo}$.

Dissolved CO_2 and CH_4 in DSDP Sediments

The content and δC^{13} of total dissolved CO_2, and the δC^{13} of methane in samples from DSDP sediments have been measured at UCLA and reported in the various volumes of the Initial Reports of the DSDP [*Presley and Kaplan, 1972; Presley et al. 1973; Claypool et al. 1973*]. In Figures 7 through 10, selected δC^{13} values of CH_4 and ΣCO_2 are plotted against depth of burial. These figures illustrate the generally parallel relation between the δC^{13} values of these dissolved carbon compounds. This relation is the principal evidence that bacterial methane production in marine sediments proceeds by the mechanism of CO_2 reduction. An alternative explanation for this relation would be carbon isotopic exchange by

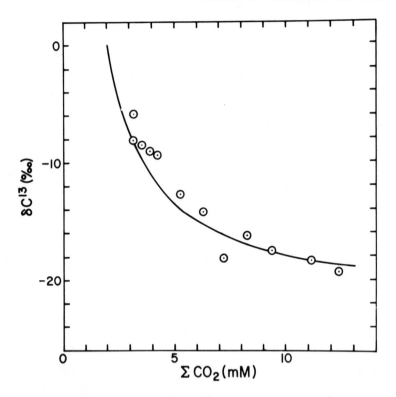

Fig. 4. Change in δC^{13} with respect to concentration of total dissolved CO_2 in the interstitial water of Santa Catalina Basin sediments [*Presley and Kaplan, 1968*]. The line is a mixing curve calculated for the addition of HCO_3^- with $\delta C^{13} = -22.5°/_{oo}$ to the original 2 mmol concentration with $\delta C^{13} \approx 0°/_{oo}$.

reversible reaction between CO_2 and CH_4 [*Bottinga, 1969*]; however, this reaction by an inorganic mechanism is so slow at low temperatures that it could not account for the effects observed in very young sediments and in laboratory experiments [*Nakai, 1961; Rosenfeld and Silverman, 1959*], and bacteria are not known to catalyze the oxidation of methane in the absence of molecular oxygen [*Stanier, et al. 1970*].

A common feature of many of the δC^{13} profiles illustrated is a δC^{13} maximum at intermediate depths and a trend toward lighter values at greater depth. This is interpreted as being due to the

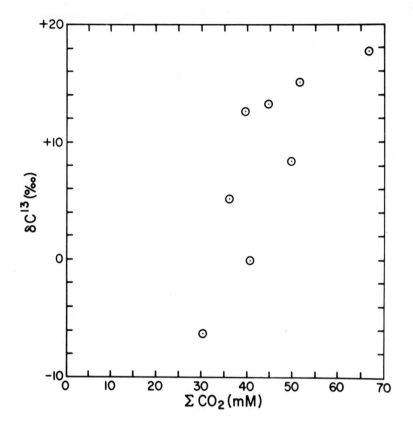

Fig. 5. Change in δC^{13} with respect to concentration of total dissolved CO_2 in the interstitial water of Saanich Inlet sediments [*Nissenbaum et al. 1972*].

addition of isotopically light (-20°/oo) CO_2 at a faster rate than CO_2 removal in methane production. The ΣCO_2 concentration also often increases over the depth interval in which this trend toward lighter δC^{13} values occurs, lending support to this interpretation. However, as ΣCO_2 concentration is sensitive to other processes, such as carbonate precipitation and dissolution, there is not always a correlation between changes in δC^{13} and concentration of ΣCO_2.

An additional implication of this trend toward lighter δC^{13} values at depth is that bacterial metabolism continues, even after burial under as much as 800 meters of sediment, and in pore waters that have been buried for times on the order of tens of millions of years.

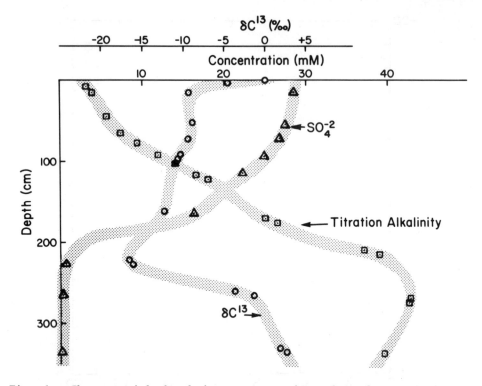

Fig. 6. Changes with depth in concentration of sulfate and titration alkalinity and the δC^{13} of total dissolved CO_2 in the interstitial water of South Guymas Basin sediments [*Goldhaber*, *1974*]. The shallowest occurrence of methane in these sediments is at 2 meters depth where trends in the plotted pore water constituents reverse.

Quantitative Estimates of Biological Methane Production

Because pressure-tight sampling equipment was not used, we have no direct quantitative measurements of the *in situ* dissolved methane concentration in the gas-rich sediments cored in the Deep Sea Drilling Project. *Hammond et al.* [*1973*] have estimated gas contents in the interstitial water of Cariaco Trench sediments (Leg 15, Hole 147) by measuring gas pocket pressures and the degree of volume expansion of the mud when brought to the surface. By this method, *in situ* contents in the range of 14 to 32 mmol/l are calculated, with no apparent correlation between amount and depth. Another independent estimate of the quantity of methane generated can be made on the basis of changes in the δC^{13} of the methane, on the assumption it has been generated by reduction of dissolved CO_2. *Nissenbaum et al.* [*1972*] related the δC^{13} of residual dissolved ΣCO_2 to the per cent CO_2 reduced by a Rayleigh distillation calculation. This same general method can be applied to DSDP sediments, where changes in $\delta C^{13}_{CH_4}$, $\delta C^{13}_{\Sigma CO_2}$, and $C_{\Sigma CO_2}$ can each be used to give an estimate of the amount of methane production, under a special set of assumptions. In order for this kind of a calculation to be valid, the required assumptions are:

1) Changes with depth in a core are equivalent to changes with time at a single depth;

2) Methane production is the only process affecting the δC^{13} and the concentration of dissolved CO_2.

In addition, the instantaneous kinetic isotope fractionation factor α must be known along with the δC^{13} and concentration of the dissolved bicarbonate at the start of the methane production process.

For the special case of closed-system CO_2 reduction (i.e., where a fixed amount of dissolved carbonate can be converted to methane via a Rayleigh distillation process) with fractionation factor α of 1.07 and δC^{13} of initial dissolved carbonate $\delta^0 = -20^\circ/_{oo}$, the relation between the cumulative methane δC^{13}, and the residual dissolved carbonate δC^{13} is shown in Figure 11, calculated from the following equations

$$\delta CO_2 = \left[\delta^0 + 1000(1 - F)^{1/\alpha - 1} \right] - 1000$$

$$\delta CH_4 = \left[\frac{\delta^0 + 1000}{F} (1 - [1 - F]^{1/\alpha}) \right] - 1000$$

where F is the fraction of the original bicarbonate converted to methane. This theoretical relation is consistent with the observed δC^{13} values of coexisting methane and dissolved carbonate in many DSDP cores, at least during the early stages of methane production. By 'early stages' of methane production, we mean the depth interval in cores where methane production is first detected and where δC^{13} of dissolved ΣCO_2 and methane become more positive, down to depths where the increase in δC^{13} stops or reverses. This is represented by the depth intervals 20 to 200 meters in hole 102 (Figure 7), 5 to 40 meters in hole 147 (Figure 8), and 100 to 300 meters in holes 174A and 180 (Figures 9 and 10). Below this depth, or past the 'early stage' of methane production, the δC^{13} of the methane no longer follows the simple closed-system Rayleigh relation, because

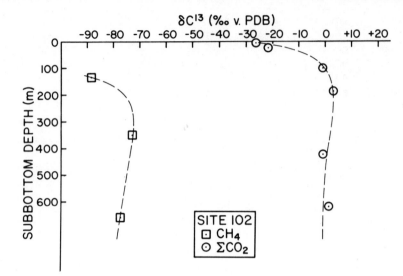

Fig. 7. Parallel changes with depth in δC^{13} of dissolved carbonate and methane in sediments of Blake-Bahama outer ridge (DSDP Site 102).

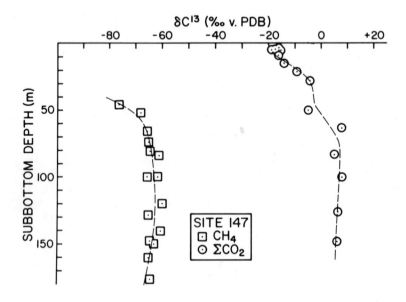

Fig. 8. Parallel changes with depth in δC^{13} of dissolved carbonate and methane in sediments of Cariaco Trench (DSDP Site 147).

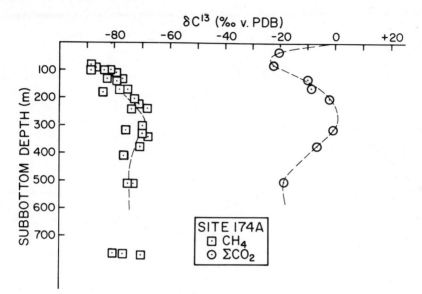

Fig. 9. Parallel changes with depth in δC^{13} of dissolved carbonate and methane in sediments of Astoria Fan (DSDP Site 174A).

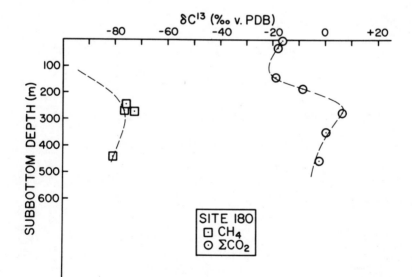

Fig. 10. Parallel changes with depth in δC^{13} of dissolved carbonate and methane in sediments of the eastern Aleutian Trench (DSDP Site 180).

the rate of addition of isotopically light ($-20°/_{oo}$) bicarbonate begins to exceed the rate of C^{12} depletion through methane production.

Figure 11 can therefore be used to estimate the amount of methane generated by CO_2 reduction. About 50% completion of a simple, single-stage Rayleigh process would be required to cause the typical methane δC^{13} shift from $-90°/_{oo}$ to $-70°/_{oo}$. The maximum dissolved ΣCO_2 concentration observed in DSDP sediments, at depths where dissolved sulfate reaches zero and methane production starts, is about 30 mmol/kg. If half of this is removed as methane, about 15 mmol/kg of dissolved methane could be generated in the pore waters during the early stages of methane production. However, dissolved ΣCO_2 may also be removed by precipitation as insoluble carbonate, or added by further decomposition of organic matter. The closed system single-stage analysis is correspondingly in error by the degree to which these processes are occurring simultaneously with methane production.

We can modify the simple single-stage Rayleigh equation to accommodate simultaneous methane production and carbonate precipitation. In the absence of significant amounts of biogenic CO_2 production, these processes would be expected to occur simultaneously because methane production tends to raise the pH and favor carbonate precipitation as shown in the following reactions

$$HCO_3^- + 8H \rightarrow CH_4 + 2H_2O + OH^-$$

$$HCO_3^- + OH^- \rightarrow CO_3^= + H_2O$$

$$Me^{++} + CO_3^= \rightarrow MeCO_3$$

$$\overline{\phantom{Me^{++} + 2HCO_3^- + 8H \rightarrow CH_4 + MeCO_3 + 3H_2O}}$$

$$Me^{++} + 2HCO_3^- + 8H \rightarrow CH_4 + MeCO_3 + 3H_2O$$

The modified form of the Rayleigh equation can be derived from the integrated equations describing parallel first-order reactions [*Daniels and Alberty, 1967*] by methods such as those outlined by *Broecker and Oversby* [*1971*]. A calculation of the relation between δC^{13} values of cumulative methane, cumulative authigenic carbonate, and residual dissolved bicarbonate for closed system simultaneous methane production and carbonate precipitation are shown in Figure 12, obtained from the following equations

$$\delta_{CO_2} = \left[(\delta^0 + 1000)(1 - F)^{\frac{1}{2}(1/\alpha - 1)} \right] - 1000$$

$$\delta_{CH_4} = \left[\left(\frac{\delta^0 + 1000}{F} \right) \frac{2}{1 + \alpha} \left(1 - (1 - F)^{\frac{1}{2}(1/\alpha + 1)} \right) \right] - 1000$$

$$\delta_{MeCO_3} = \left[\left(\frac{\delta^0 + 1000}{F} \right) \frac{2}{1/\alpha + 1} \left(1 - (1 - F)^{\frac{1}{2}(1/\alpha + 1)} \right) \right] - 1000$$

The initial dissolved bicarbonate δ^0 is $-20°/_{oo}$, the fractionation factor α for methane production is 1.07, and no fractionation is involved in carbonate precipitation. In this case, about 67% removal F of dissolved ΣCO_2 would shift the δC^{13} of the methane from $-83°/_{oo}$ to $-70°/_{oo}$. Again, taking the initial dissolved ΣCO_2 concentration as 30 mmol/kg at the beginning of the process, the reactions lead to formation of 10 mmol/kg each as methane and authigenic carbonate deposit.

The closed system single-stage process portrayed in Figure 11 predicts that for 50% ΣCO_2 reduction, the residual dissolved ΣCO_2 will decrease in concentration from 30 to 15 mmol/kg and increase in δC^{13} from -20 to $+25°/_{oo}$, whereas the closed system reaction process shown in Figure 12 predicts a concentration decrease of from 30 to 10 mmol/kg and a δC^{13} increase of from -20 to about $+10°/_{oo}$. The most commonly observed changes in dissolved bicarbonate in DSDP sediments during the early stage of methane production are from 30 to 5 mmol/kg and δC^{13} change from about -23 to about $+5°/_{oo}$. This corresponds more closely to the changes predicted by the calculation of simultaneous methane production and carbonate precipitation in Figure 12, suggesting that 10 mmol/kg for early stage methane production may be a better estimate. This calculation neglects CO_2 generation during the early stage of methane generation.

Many DSDP cores show a deeper second maxima in dissolved ΣCO_2 of alkalinity (for example, at 126 meters in hole 147, at 500 meters in hole 104, at 300 meters in hole 174, at 330 meters in hole 178, at 250 meters in hole 180, and at 120 meters in hole 181). This late stage CO_2 production process is clearly unrelated to sulfate reduction, and may represent a low-temperature thermocatalytic decarboxylation of organic acid functional groups, similar to that suggested by *Johns and Shimoyama* [1972]. This late stage CO_2 production process appears to be controlled by the age and the inferred temperature of the sediments, arguing against a biological mechanism. The CO_2 that is generated in this process is isotopically light ($\delta C^{13} = -20°/_{oo}$) and apparently serves as a carbon source for further methane production, as indicated by the shift toward lighter δC^{13} values in deeper samples of methane. For example, in hole 174A (Figure 9), the δC^{13} of methane at 250 meters is about $-70°/_{oo}$,

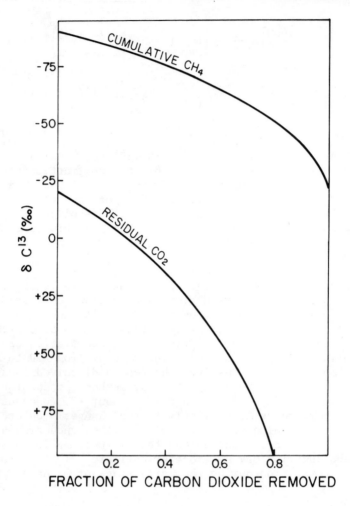

FRACTION OF CARBON DIOXIDE REMOVED

Fig. 11. Carbon isotope fractionation in closed-system single-stage Rayleigh processes.

and at 500 meters, it is about $-75°/_{oo}$. At the same depths in this hole, the dissolved bicarbonate δC^{13} is $0°/_{oo}$ and $-20°/_{oo}$, respectively. It appears that the cumulative methane δC^{13} has shifted from $-70°/_{oo}$ to $-75°/_{oo}$ by addition of isotopically lighter carbon associated with the late stage CO_2 production process. If the concentration of dissolved methane is 10 mmol/kg when late-stage CO_2 production begins, and the δC^{13} of methane added after this point averages $-80°/_{oo}$, another 10 mmol/kg would be required to shift the δC^{13} of the methane back to $-75°/_{oo}$. The above suggests

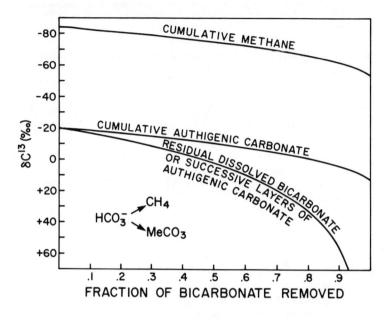

δC^{13} (‰)

CUMULATIVE METHANE

CUMULATIVE AUTHIGENIC CARBONATE

RESIDUAL DISSOLVED BICARBONATE OR SUCCESSIVE LAYERS OF AUTHIGENIC CARBONATE

$HCO_3^- \rightarrow CH_4$
$HCO_3^- \rightarrow MeCO_3$

FRACTION OF BICARBONATE REMOVED

Fig. 12. Carbon isotope fractionation in closed-system, branching Rayleigh process.

that typically a core taken in rapidly-depositing anoxic marine sediment with about 0.5% organic carbon may generate $\gtrsim$ 20 mmol/kg methane at or below sediment depths equal to 30°C. This is a minimum estimate of *in situ* biological methane generation, and does not take into account any upward migration of methane generated from greater depths in the sediment colume. Thermophilic methanogenic bacteria have recently been isolated from sewage sludge [*Zeikus and Wolfe, 1972*] so that biological methane production could conceivably occur down to burial depths equivalent to 65-75°C. The activation energy for thermocatalytic methane generation is about 40.3 kcal/mole [*Hoering and Abelson, 1963*] compared to 36 kcal/mole for decarboxylation [*Johns and Shimoyama, 1972*]. Therefore, nonbiological methane generation at detectable rates should only begin at temperatures above 50°C, which would be reached in deep water marine sediments at burial depths of about 1500 meters, assuming an average geothermal gradient of 3.0°C/100 meters. The above discussion also assumes that biological methane production from carbon sources other than CO_2, such as acetate, is insignificant in marine sediments. This assumption may be incorrect, in which case CH_4 in concentrations > 20 mmol/kg interstitial water could form at relatively shallow depths.

DIAGENETIC EFFECTS ASSOCIATED WITH METHANE OCCURRENCE IN MARINE SEDIMENT

In order to evaluate possible sedimentological effects produced during methane generation, it is necessary to know something about the state of the gas under the *in situ* sediment conditions. As methane is by far the most abundant gas in almost all samples analyzed [*Claypool et al. 1973*], and because the range of possible sedimentary temperatures is above the critical point of methane, the possible states in which gas can exist in sediments are: methane in solution in the pore water, methane-saturated pore water plus a free gas phase, methane-saturated pore water plus a solid clathrate phase (i.e., methane hydrate); and methane hydrate plus a free gas phase.

Gas Hydrate

Methane and many other substances that are normally gaseous under earth surface conditions combine with water to form crystalline, ice-like clathrate compounds under certain conditions of high pressure and low temperature [see *Miller, 1974; Hand et al. 1974;* and *Hitchon, 1974*]. The stability relationships of these compounds have been thoroughly studied, and a schematic pressure-composition phase diagram for the methane-water system at 25°C adopted from *Kobayashi and Katz* [*1949*], using the experimental values for the solubility of methane reported by *Culberson and McKetta* [*1951*] and *Makogon et al.* [*1972*] is shown in Figure 13. At this temperature, 25°C, the requirements for methane hydrate ($CH_4 \cdot 5-3/4H_2O$) stability are: pressure in excess of 462 atm, and a methane concentration that exceeds about 0.13 mole %, or about 72 mmol/kg. For methane in sea water, the critical concentration for hydrate stability would be about 20% less, or around 58 mmol/kg, assuming that the decrease in methane solubility caused by 35°/₀₀ salinity is comparable to that for nitrogen and argon, as discussed by *Weiss* [*1970*].

The pressure-temperature phase relations for methane hydrate are illustrated in Figure 14. This diagram is a P-T section of the three-dimensional pressure-temperature-composition stability field at a bulk composition in the range 0.4 to 14.8 mole % methane in water. Because this amount of methane is in excess of that which can be dissolved in the water under the conditions represented in Figure 14, methane will also be present as a free gas phase or as the solid methane hydrate. Gas will be in equilibrium with the hydrate under the various combinations of pressure and temperature indicated by the diagonal lines of Figure 14. This series of univariant equilibrium lines illustrates: the depression of the freezing point of methane hydrate by dissolved salts of sea water salinity, and the stabilization at lower pressures of methane hydrate by small amounts of gases, such as CO_2 and H_2S, which form

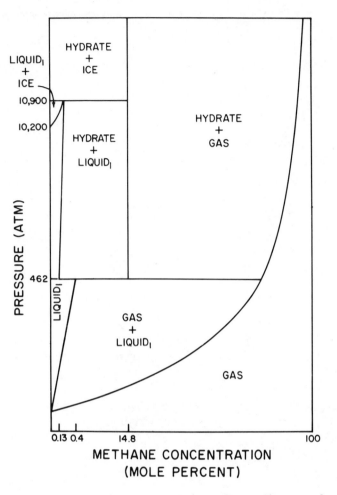

Fig. 13. Schematic pressure-composition phase diagram for H_2O-CH_4 system at 25°C.

hydrates at lower pressures. These effects are of similar magnitude in opposite directions for gases in marine sediments. Therefore, as the effects of salinity and minor or trace amounts of other gases approximately cancel each other, the pressure temperature stability conditions for gas hydrates in marine sediments are about the same as in the pure methane-water system.

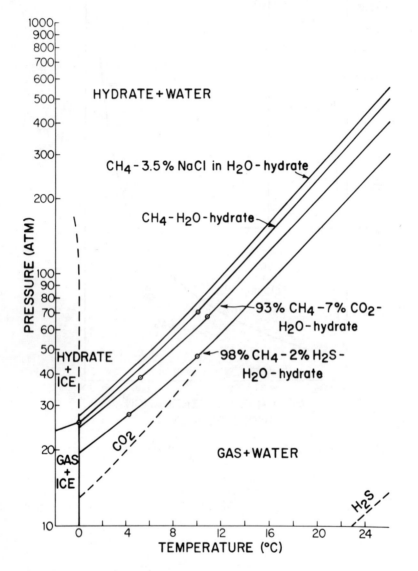

Fig. 14. Pressure-temperature phase diagram for H_2O gas systems.

Changes in Methane Solubility with Depth

One implication of the pressure-composition phase diagram for the methane-water system (Figure 13) is that under a given set of conditions, a minimum methane concentration is required to stabilize either a free gas phase or a solid clathrate hydrate phase. Most deep sea sediments do not contain sufficient organic matter to support the activity of methane-producing bacteria. Those that do are limited in the amount of methane that can be generated. In order to evaluate the probable limits of hydrated methane formation in deep sea sediments, it is necessary to compare the estimates of methane generated to the quantities of methane that can be dissolved in the pore water. It is for this purpose that Figure 15 has been prepared, illustrating the approximate relation between methane solubility and depth of burial in marine sediments for a series of overlying water depths: 500 meters, 1000 meters, 2000 meters, 3000 meters, and 4000 meters. In this diagram, the combined effects of pressure and temperature on methane solubility are plotted as a function of depth of burial.

The *in situ* pressure and temperature of sediments at any depth of burial are estimated from:

1) The total height of the overlying water column from the point of burial to sea level, assuming a hydrostatic pressure gradient of 0.1 atm/m;

2) The depth of burial beneath the sediment-water interface, assuming a thermal gradient of 0.035°C/m and a bottom water temperature of 2°C (5°C at 500 meters water depth).

The equilibrium solubility of methane in water under a given set of P-T conditions is estimated by interpolation from the experimental data of *Culberson and McKetta* [1951], with a linear 20% reduction for salinity correction. In addition, these methane solubility measurements were extrapolated into the P-T region of methane hydrate stability, and reduced by two-thirds in accordance with the observation of *Makogon et al.* [1972] that "at T and P corresponding to the hydrate equilibrium, the solubility of methane in water decreases abruptly, by a factor of 3 to 5". This information recorded in Figure 15 can be used to predict the state of gas occurrence in marine sediments, given the following information: water depth, depth of burial, and dissolved methane concentration.

Under conditions such as those existing on the continental borderland off the west coast of North America, the minimum water depth required to stabilize methane hydrate is about 500 meters, assuming a bottom water temperature of 5°C. (It should be noted that at high latitudes, bottom temperatures as low as -2°C are possible and methane hydrate would be stabilized at correspondingly lower

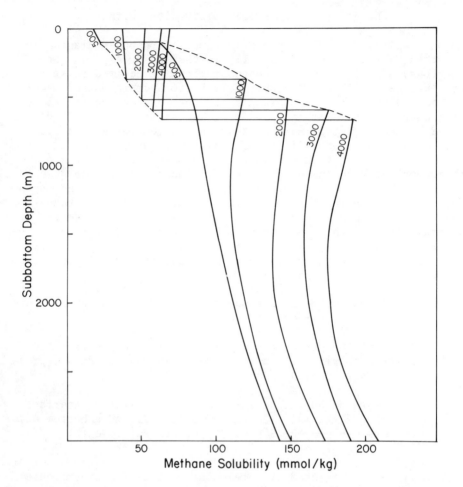

Fig. 15. Methane solubility with respect to depth of burial in deep-sea sediments. Curves represent pore water methane saturation in sediments under a specified water depth (as labeled in meters). Dashed lines indicate methane solubility with respect to water depth at the maximum burial depths at which hydrates are stable (left dashed line) and at the minimum depth at which a free gas phase can exist (right dashed line).

hydrostatic pressures or water depth). For typical conditions in 500 meters of water, methane hydrate is stable in the upper 100 meters of sediment, if the methane concentration exceeds the saturation value, plotted as 18 to 22 mmol/kg, as shown by the curve labeled 500 in Figure 15. Below a burial depth of 100 meters, methane hydrate is not stable, and the solubility of methane abruptly increases

to about 58 mmol/kg. This contrasts with the situation in sediments
covered by 4000 meters of water, and shown by the curve labeled 4000
in Figure 15. Under these conditions, methane hydrate is stable in
the sediments down to a burial depth of 670 meters, when the amount
of methane present exceeds the saturation concentration shown in
Figure 15 as 63 to 69 mmol/kg. Below 670 meters, the methane solubility in the interstitial water abruptly increases to 190 mmol/kg.

The dashed line at the left of Figure 15 represents methane
saturation at the maximum burial depth of hydrate stability in marine sediments, assuming reasonable P-T conditions for a continuous
range of water depths from 500 to 4000 meters. At shallower burial
depths, methane in excess of saturation combines with the interstitial water to form the solid gas hydrate until a concentration of
14.8 mole % methane $CH_4 \cdot 5\text{-}3/4H_2O$ is reached, at which point all of
the available water is bound up in the clathrate structure. Methane
in excess of this amount (if it is attainable under natural conditions), will remain in the gas phase. For deeper burial, where sediment temperatures exceed hydrate decomposition temperatures, the
structure of liquid water apparently changes, and the amount of methane that can be dissolved in the interstitial water increases by a
factor of 3 or more [*Makogon et al. 1972*]. The dashed line at the
upper right of Figure 15 indicates this situation when methane solubility is approached from beneath the zone of gas hydrate stability.
In this case, the upper dashed line connects points of maximum methane solubility at the minimum burial depths for stability of a free
gas phase. Solid horizontal lines that connect the two dashed lines
indicate only the abrupt change in methane solubility occurring at
the burial depth of the hydrate-gas phase transition.

Figure 15 also shows that for deep sea (> 1000 meters) sediments
beneath the zone of gas hydrate stability, the solubility of methane
in interstitial water decreases with increasing depth, and reaches a
solubility minimum between 1 and 2 km burial depth. Sediments undergoing burial must travel down this gradient of decreasing methane
solubility. If conditions of sedimentation and diagenesis are such
that methane generation saturates the interstitial waters at some
depth above the solubility minimum, the interstitial waters will become supersaturated with increased depth of burial, and eventually
give rise to a free gas phase. In deep water sediments, this compressed gas could migrate back up the pressure gradient until: the
gas encounters sediment in which the interstitial water is not saturated with gas where it will dissolve, or the gas encounters sediments that are in the temperature stability field of methane hydrate,
where it may combine with water to form the gas hydrate. In this
manner, large quantities of methane gas could be concentrated in the
upper kilometer or so of thick accumulations of deep-sea sediment.

The temperature range of the methane solubility minimum (40 to
60°C) is high enough that thermocatalytic generation of hydrocarbons

may occur and contribute to pore water methane saturation at shallower depths. Indirect evidence for the vertical migration of gas was present in DSDP core 106 from Leg 11, in the form of vertical fractures and 'burrow-like' structures lined with siderite [*Lancelot and Ewing, 1973*].

Evidence for Gas Hydrates in DSDP Sediments

Methane-generating sediments of sufficient thickness to enable the above mechanism to concentrate significant amounts of gas have not been commonly sampled in the deep sea. However, at least one such vast accumulation of sediments exists in the Blake-Bahama outer ridge system on the continental rise off the east coast of North America. These sediments were cored on Leg 11 of the DSDP. The shipboard scientific party [*Hollister et al. 1973*] made the following observations concerning the sediments of the Blake-Bahama outer ridge (Sites 102, 103, 104):

1) The sediments contained substantial amounts of methane gas;

2) A prominent seismic reflector occurs at about 0.61 to 0.62 sec (sub-bottom two-way travel time);

3) A change in drilling penetration rate occurred at about 615 to 620 meters at sites 102 and 104;

4) A lithologic change, from soft hemipelagic to indurated hemipelagic mud containing siderite or ankerite modules or layers, occurred at the same depth as the break in drilling rate;

5) Both the seismic reflector and the lithologic change cross time surfaces, and parallel sea floor topography.

On the basis of the most obvious correlation, that the seismic reflector, the drilling break, and the lithologic change all occur at the same depth (610 to 620 meters), the sediments transmit seismic waves at an anomalously fast average velocity, 2.2 km/sec, compared to normal velocities of 1.7 to 1.9 km/sec in sediments with similar porosity and grain size [*Stoll, 1974, and Bryan, 1974*]. *Stoll et al.* [*1971*] investigated the effects of gas on the acoustic wave velocity of sediments. They found that the seismic velocity in water-saturated sand increased from 1.7 to more than 2.5 km/sec under conditions that supported formation of methane hydrate.

Within a given region in the deep ocean, isobars parallel the sea surface, and isotherms parallel the sea floor topography; thus the position of the seismic reflector and the lithologic change appear to be controlled mainly by temperature. Because of the relatively flat slope of the P-T gradients within the sediments, as shown for holes 106, 102, 88, 185, and 147 in Figure 16, the intersection with the phase boundary is mostly a function of change in temperature (or depth of burial). This is the principal line of evidence that links the methane gas-solid phase change to the observed time-transgressive seismic and lithologic effects.

If the sediments in the depth range of 100 to 600 meters in the region of the Blake-Bahama outer ridge contain hydrated methane, as indicated by the anomalously fast acoustic velocities, the pressure and temperature conditions of these sediments must be consistent with the known P-T stability field of methane hydrate. In Figure 16, depth scales have been superimposed on the pressure and temperature axes of the P-T phase diagram, and estimated P-T gradients plotted for some of the DSDP sites in which gassy sediments were encountered. Shown also is the temperature at the depth (based on the assumed geothermal gradient of 0.03°C/m) of the bottom-simulating seismic reflector in the Blake-Bahama outer ridge sediments (site 102) and in the Bering Sea (site 185). These reflectors could be made to coincide exactly with the methane hydrate phase boundary by small adjustments in the assumed geothermal gradients; a 0.035°C/m gradient would be required in site 102, for example. P-T gradients from selected other locations are also included in Figure 16. The gas in sediments of site 88 in the Gulf of Mexico should be present *in situ* as dissolved methane plus methane hydrate if the amount of gas present exceeds saturation. In contrast, the sediments of the Cariaco Trench have abundant methane, which should all be present as a dissolved or as a gas phase, as the hydrate is not stable at the relatively high temperature of these sediments. The P-T gradient of the gas-containing sediments of site 106, the lower continental rise southeast of New York, intersect the hydrate-gas phase transition at about 820 meters, assuming 0.03°C/m geothermal gradient. There is a 'moderately distinct' reflector in the seismic profiler record at 7.1 sec, or about 1.2 sec sub-bottom reflection time. At this same location, a minor break in drilling rate occurred at 920 meters, and a major break at 1020 meters. A lithologic change from hemipelagic mud to siderite-rich silicified claystone was confirmed by coring, slightly above 935 meters. The seismic reflector and the major drilling break have been interpreted to correlate with each other and with Horizon A, the widespread seismic reflector off the coast of eastern North America [*Hollister et al. 1973*]. If, in fact, the geothermal gradient is slightly lower than average and is near 0.025°C/m, it would place the gas-hydrate phase transition at about 920 meters, and the relations among gas-rich sediments, lithologic changes, and seismic reflectors would be closely analogous to the situation in the Blake-Bahama Ridge sediments, except for anomalous acoustic velocities.

The Nature of the Reflector

The exact nature of the bottom-simulating reflector is not

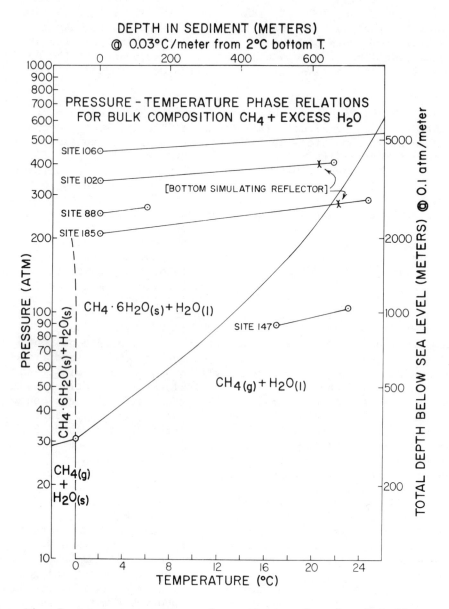

Fig. 16. Pressure-temperature phase diagram for the H_2O-CH_4 system at bulk composition in the range 0.4 to 14 mole % CH_4. Subsea depth and sub-bottom depth scales have been superimposed on P-T axes, respectively, assuming hydrostatic pressure and normal geothermal gradients.

known, but at least three explanations have been proposed. *Lancelot* [*1971*] and *Lancelot and Ewing* [*1973*] related the reflector to a zone of siderite or ankerite nodules, although they pointed out in a later publication that correspondence between the drilling break, the reflector, and a well-defined lithologic change could not be definitely established. *Ewing and Hollister* [*1973*] suggested two other hypotheses to account for the reflector: "(1) the reflector may correspond to the isotherm that separates a gas environment from a hydrate environment, and (2) the hydrate/gas isotherm may be below reflector Y, and the reflector is due to a concentration of hydrated methane produced by gas migrating upward from the deeper layers into the hydrate sustaining environment."

Subsequent drilling in the Bering Sea-Aleutian Arc region on Leg 19 of the DSDP has provided additional evidence on the nature of the seismic reflector. This bottom-simulating reflector occurs at 0.72 sec in hole 184 and at 0.82 sec in hole 185. In both holes, a lithologic change was observed that would coincide with the reflector at velocities of 1.68 and 1.65 km/sec in the overlying sediments. The lithologic discontinuity separates a siltstone-mudstone sequence below from unlithified hemipelagic muds above. Gas was not detected in hole 184, and was present only at depth greater than 660 meters in hole 185. The seismic velocities are consistent with the absence of gas hydrates in the unlithified section, compared with the anomalously fast velocities and the inferred presence of gas hydrates in Leg 11 sediments (holes 102 and 104). On the basis of the Leg 19 results, it seems the most probable cause of the seismic reflector is the lithologic discontinuity. The presence of a free gas phase, a section of frozen sediment, or a zone of carbonate nodules is probably not the primary cause of the reflector, although any one, or each, may be associated with the lithologic change.

However, the depth of the lithologic change is, in all cases, consistent with the inferred depth of the critical isotherm for the decomposition of gas hydrate under the prevailing pressure conditions, and suggests at least an indirect link to gas diagenesis. The sedimentological nature of the lithologic change is that of an abrupt transition from unlithified hemipelagic mud to indurated claystone. The lithification process apparently involves the dissolution of calcareous and siliceous microfossils, along with unstable detrital minerals, and the reprecipitation of carbonate, silica, and clay mineral cement. Such a lithologic transition in gassy sediments might be accounted for by the following sequence of events:

1) Rapid methane production causing a rise in pH, dissolution of siliceous skeletal material, and precipitation of carbonate;

2) Temperature-controlled release of molecular CO_2 from the organic matter into the pore water (i.e., decarboxylation) causing a lowering of pH, precipitation of silica, dissolution of more readily soluble carbonate, and/or silicate components of the sediment, resulting in an increase in bicarbonate concentration;

3) Renewed methane production, increase in pH, and precipitation of authigenic carbonate.

There are at least two ways that reprecipitation of authigenic minerals could be influenced by the depth of the gas hydrate phase transition in the sediment. First, the presence of a free gas phase beneath the hydrated zone could create open space in the sediment within which precipitation could take place. There is ample evidence for this in the description of the Leg 11 sediments [*Lancelot and Ewing, 1973*]. Examples include the observation in the deeper parts of the holes 102, 104 and 106 of siderite-filled veinlets, fractures, 'burrows', globules, concretions, and layers. In addition, [*Lancelot et al. 1973*] note that quartz and feldspars are occasionally found "concentrated in small white spots and streaks that resemble burrow fillings". Second, if CO_2 is liberated in the hydrated zone, it may be tied up as a clathrate component, inhibiting reaction with the sediment, until a burial depth (i.e., temperature) is reached where the gas hydrate decomposes. This could explain why calcareous and siliceous microfossils are preserved in the theoretical stability depth range of gas hydrates and destroyed below it.

CONCLUSIONS

The presence of methane has been demonstrated in deep sea, as well as shallow marine environments. A necessary condition for methane formation is rapid deposition of sediment with sufficient organic carbon ($\geq 0.5\%$) to allow anoxic conditions to be established. Methane production begins in the sediment column after the dissolved sulfate has been reduced. All the pathways for methane formation are not known, but it is believed that reduction of CO_2 by biologically-produced hydrogen is the single most important mechanism.

Under proper conditions of P and T, methane will form gas hydrates that may crystallize in the sediment. Acoustic reflection studies have given indirect evidence for anomalous layering in the sediment that cuts across time-stratigraphic boundaries. Although depth and temperature conditions on the abyssal sea floor are favorable for methane hydrate formation, the evidence presently suggests that sufficient methane is present only in the thicker and more rapidly-deposited deep sea sediments. Estimates made from δC^{13} data based on Rayleigh models suggest that only 20-30 mmol/kg methane are formed biologically. A concentration mechanism is possible because methane solubility initially decreases with increasing temperature or depth of burial, and this may lead to upward migration from deeper gas-saturated sediments.

Gas diagenesis may be related to seismic anomalies in deep sea sediments. In addition to the effects of free or hydrated gas on acoustic wave transmission, methane generation may also be related to lithification of muds beneath the zone of gas hydrate stability.

ACKNOWLEDGMENTS

This study was supported in part by the Atomic Energy Commission, under Contract No. AT(04-3)34 P.A. 134.

Publication No. 1331, Institute of Geophysics and Planetary Physics, University of California, Los Angeles.

REFERENCES

Bottinga, Y., Calculated fractionation factors for carbon and hydrogen isotope exchange in the system calcite-carbon dioxide-graphite-methane-hydrogen-water vapor, *Geochim. et Cosmochim. Acta, 33,* 49, 1969.

Broecker, W. S., and V. M. Oversby, *Chemical Equilibrium in the Earth,* 318 pp., McGraw-Hill, 1971.

Bryan, G. M., *In situ* indications of gas hydrate, in *Natural Gases in Marine Sediments* edited by I. R. Kaplan, pp. 299-308, Plenum Press, New York, 1974.

Bryant, M. P., Rumer methanogenic bacteria, in *Physiology of Digestion in the Ruminant* edited by R. W. Dougherty *et al.,* pp. 411-418, Butterworths, Washington, 1965.

Bryant, M. P., E. A. Wolin, M. J. Wolin, and R. S. Wolfe, Methanobacillus omelianskii, a symbiotic association of two species of bacteria, *Arch. Mikrobiol., 59,* 20, 1967.

Claypool, G. E., B. J. Presley, and I. R. Kaplan, in *Initial Reports of the Deep Sea Drilling Project,* vol. 19, pp. 879-884, U. S. Government Printing Office, Washington, 1973.

Columbo, U., F. Gazzarrini, R. Gonfiantini, G. Sironi, and E. Tongiorgi, Measurements of C^{13}/C^{12} isotope ratios on Italian natural gases and their interpretation, in *Advances in Organic Geochemistry 1964,* edited by G. D. Hobson and M. C. Louis, pp. 279-292, Pergamon Press, New York, 1966.

Craig, H., The geochemistry of stable carbon isotopes, *Geochim. et Cosmochim. Acta, 3,* 53, 1953.

Culberson, O. L., and J. J. McKetta, Solubility of methane in water at pressures to 10,000 psi, *J. Petrol. Technol.*, *3*, (*Trans*. AIME), 223, 1951.

Daniels, F., and R. A. Alberty, *Physical Chemistry*, 767 pp., John Wiley & Sons, Inc., New York, 1967.

Emery, K. O., and D. Hoggan, Gases in marine sediments, *Amer. Ass. Petrol. Geol. Bull.*, *42*, 2174, 1958.

Ewing, J. I., and C. D. Hollister, in *Initial Reports of the Deep Sea Drilling Project*, vol. 11, pp. 951-976, U. S. Government Printing Office, Washington, 1973.

Goldhaber, M. B., Equilibrium and dynamic aspects of the marine geochemistry of sulfur, Ph.D. thesis, Univ. of California, Los Angeles, 1974.

Hammond, D. E., R. M. Horowitz, and W. S. Broecker, in *Initial Reports of the Deep Sea Drilling Project*, vol. 20, U. S. Government Printing Office, Washington, 1973.

Hand, J. H., D. L. Katz, and V. K. Verma, Review of gas hydrates with implications for ocean sediments, in *Natural Gases in Marine Sediments*, edited by I. R. Kaplan, pp. 179-194, Plenum Press, New York, 1974.

Hardin, G., *Biology Its Principles and Implications*, 77 pp., W. H. Freeman and Co., 1966.

Hitchon, B., Occurrence of natural gas hydrates in sedimentary basins, in *Natural Gases in Marine Sediments* edited by I. R. Kaplan, pp. 195-225, Plenum Press, New York, 1974.

Hoering, T. C., and P. H. Abelson, Hydrocarbons from kerogen, Annual Rept., Director Geophys. Lab., *Carnegie Inst. Wash. Yr. Bk.*, *62*, 229, 1963.

Hoering, T. C., Organic acids from the oxidation of recent sediment, *Carnegie Inst. Wash. Yr. Bk.*, *66*, 515, 1968.

Hollister, D. C., J. I. Ewing *et al.*, in *Initial Reports of the Deep Sea Drilling Project*, vol. 11, pp. 135-218, U. S. Government Printing Office, Washington, 1973.

Iannotti, E. L., D. Kafkewitz, M. J. Wolin, and M. P. Bryant, Glucose fermentation products of *Ruminococcus albus* grown in continuous culture with *Vibrio succinogenes:* Changes caused by interspecies transfer of H_2, *J. Bacteriol.*, *114*, 1231, 1973.

Johns, W. D., and A. Shimoyama, Clay minerals and petroleum-forming reactions during burial and diagenesis, *Amer. Ass. Petrol. Geol. Bull.*, *56*, 2160, 1972.

Kobayashi, R., and D. L. Katz, Methane hydrate at high pressure, *Petrol. Technol.*, *1*, (Trans. AIME), 66, 1949.

Koyama, T., Gaseous metabolism in lake sediments and paddy soils, in *Advances in Organic Geochemistry, 1962,* edited by U. Columbo and G. D. Hobson, pp. 363-375, The Macmillan Co., New York, 1964.

Lancelot, Y., Carbonate diagenesis in the gas-rich Tertiary sediments from the Atlantic North American Basin (Abst.), in *8th Int. Sedimentol. Cong.,* Heidelberg, 1971.

Lancelot, Y., and J. I. Ewing, in *Initial Reports of the Deep Sea Drilling Project,* vol. 11, pp. 791-800, U. S. Government Printing Office, Washington, 1973.

Makogon, Yu. F., V. I. Tsarev, and N. V. Cherskiy, Formation of large natural gas fields in zones of permanently low temperatures, *Dokl. Akad. Nauk SSSR (Earth Sci.),* English Transl., *205,* 215, 1972.

Manheim, F. T., K. M. Chan, and F. L. Sayles, in *Initial Reports of the Deep Sea Drilling Project,* vol. 5, pp. 501-512, U. S. Government Printing Office, Washington, 1970.

Manheim, F. T., F. L. Sayles, and L. S. Waterman, in *Initial Reports of the Deep Sea Drilling Project,* vol. 10, pp. 615-623, U. S. Government Printing Office, Washington, 1973.

Mechalas, B. J., Pathways and environmental requirements for biogenic gas production in the ocean, in *Natural Gases in Marine Sediments,* edited by I. R. Kaplan, pp. 11-25, Plenum Press, New York, 1974.

Miller, S. L., The nature and occurrence of clathrate hydrates, in *Natural Gases in Marine Sediments,* edited by I. R. Kaplan, pp. 151-177, Plenum Press, New York, 1974.

Nakai, N., Geochemical studies on the formation of natural gases, Ph.D. thesis, Nagoya Univ., Japan, 1961.

Nissenbaum, A., B. J. Presley, and I. R. Kaplan, Early diagenesis in a reducing fjord, Saanich Inlet, British Columbia - I. Chemical and isotopic changes in major components of interstitial water, *Geochim. et Cosmochim. Acta, 36,* 1007, 1972.

Oana, S., and E. S. Deevey, Carbon-13 in lake waters, and its possible bearing on paleolimnology, *Amer. J. Sci., 258-A,* 253, 1960.

Postgate, J. R., Recent advances in the study of the sulfate-reducing bacteria, *Bacteriol. Rev., 29,* 425, 1965.

Presley, B. J., J. Culp, C. Petrowski, and I. R. Kaplan, in *Initial Reports of the Deep Sea Drilling Project,* vol. 11, pp. 805-810, U. S. Government Printing Office, Washington, 1973.

Presley, B. J., and I. R. Kaplan, Changes in dissolved sulfate, calcium and carbonate from interstitial water of near shore sediments, *Geochim. et Cosmochim. Acta, 32,* 1037, 1968.

Presley, B. J., and I. R. Kaplan, *Initial Reports of the Deep Sea Drilling Project*, vol 4, pp. 415-430, U. S. Government Printing Office, Washington, 1970.

Presley, B. J., and I. R. Kaplan, in *Initial Reports of the Deep Sea Drilling Project*, vol. 11, pp. 1009-1012, U. S. Government Printing Office, Washington, 1972.

Reeburgh, W. S., Observations of gases in Chesapeake Bay sediments, *Limnol. Oceanogr.*, *14*, 368, 1969.

Rosenfeld, W. D., and S. R. Silverman, Carbon isotope fractionation in bacterial production of methane, *Science*, *130*, 1658, 1959.

Sackett, W. M., S. Nakaparksin, and D. Dalrymple, Carbon isotope effects in methane production by thermal cracking, in *Advances in Organic Geochemistry*, *1966*, edited by G. D. Hobson and G. G. Speers, pp. 37-53, Pergamon Press, New York, 1968.

Sayles, F. L., and F. T. Manheim, in *Initial Reports of the Deep Sea Drilling Project*, vol. 7, pp. 871-882, U. S. Government Printing Office, Washington, 1971.

Sayles, F. L., F. T. Manheim, and K. M. Chen, in *Initial Reports of the Deep Sea Drilling Project*, vol. 4, pp. 401-414, U. S. Government Printing Office, Washington, 1970.

Sayles, F. L., F. T. Manheim, and L. W. Waterman, in *Initial Reports of the Deep Sea Drilling Project*, vol. 11, pp. 997-1008, U. S. Government Printing Office, Washington, 1972.

Sayles, F. L., F. T. Manheim, and L. S. Waterman, in *Initial Reports of the Deep Sea Drilling Project*, vol. 12, pp. 801-808, U. S. Government Printing Office, Washington, 1973a.

Sayles, F. L., L. S. Waterman, and F. T. Manheim, in *Initial Reports of the Deep Sea Drilling Project*, vol. 19, pp. 871-874, U. S. Government Printing Office, Washington, 1973b.

Sholkovitz, E., Interstitial water chemistry of the Santa Barbara Basin sediments, *Geochim. et Cosmochim. Acta*, *37*, 2043, 1973.

Smith, P. H., and R. A. Mah, Kinetics of acetate metabolism during sludge digension, *Appl. Microbiol.*, *14*, 368, 1966.

Stanier, R. Y., M. Doudoroff, and E. A. Adelberg, *The Microbial World*, 873 pp., Prentice-Hall, Inc., Englewood Cliffs, New Jersey, 1970.

Steggerda, F. R., and J. F. Dimmick, Effects of bean diets on concentration of carbon dioxide in flatus, *Amer. J. Clin. Nutr.*, *19*, 120, 1966.

Stoll, R. E., Effects of gas hydrates in sediments, in *Natural Gases in Marine Sediments*, edited by I. R. Kaplan, pp. 235-248, Plenum Press, New York, 1974.

Stoll, R. D., J. Ewing and G. M. Bryan, Anomalous wave velocities in sediments containing gas hydrates, *J. Geophys. Res., 76,* 2090, 1971.

Toerien, D. F., and W. H. J. Hattingh, The microbiology of anaerobic digestion, *Water Res., 3,* 385, 1969.

Tuttle, J. H. and H. W. Jannasch, Dissimilatory reduction of inorganic sulfur by facultatively anaerobic marine bacteria. *J. Bact., 115,* 732, 1973.

Vaccaro, R. F., Inorganic nitrogen in sea water, in *Chemical Oceanography,* edited by J. P. Riley and G. Skirrow, pp. 365-408, Academic Press, New York, 1965.

Waterman, L. S., F. L. Sayles, and F. T. Manheim, in *Initial Reports of the Deep Sea Drilling Project,* vol. 18, pp. 1001-1012, U. S. Government Printing Office, Washington, 1973.

Weiss, R. F., The solubility of nitrogen, oxygen and argon in water and sea water, *Deep Sea Res., 17,* 721, 1970.

Whelan, T., Methane and carbon dioxide in coastal marsh sediments, in *Natural Gases in Marine Sediments,* edited by I. R. Kaplan, pp. 47-61, Plenum Press, New York, 1974.

Wolfe, R. S., Microbial formation of methane, *Adv. Microbial. Physiol. 6,* 107, 1971.

Zeikus, J. G., and R. S. Wolfe, *Methanobacterium thermoautotrophicus* sp. n., an anaerobic, autotrophic, extreme thermophile, *J. Bacteriol., 109,* 707, 1972.

Zobell, C. E., and S. C. Rittenberg, Sulfate-reducing bacteria in marine sediments, *J. Marine Res., 7,* 602, 1948.

GEOTHERMAL GASES

Graeme L. Lyon

Institute of Nuclear Sciences
Department of Scientific and Industrial Research
Lower Hutt, New Zealand

ABSTRACT

The major components of geothermal gases are steam and carbon dioxide. At high temperatures, chemical and isotopic equilibrium will be established, but as the gases cool, equilibrium conditions are not maintained. Interaction of gases with wet sediments will result in a gas mixture consisting of methane, hydrogen, nitrogen and rare gases.

INTRODUCTION

Ocean-related volcanism occurs at plate margins in the circum-Pacific belt, mid-Atlantic ridge and other regions, and in areas apparently unrelated to plate margins such as the Hawaiian Islands. Submarine volcanism is likely to be widespread. In regions of low grade volcanic activity, geothermal gases may be released into the sediments on the sea floor or through the sediments into the ocean. It is thus of some interest to compare geothermally derived gases with the other natural gases discussed in this symposium.

THE COMPOSITION OF THERMAL WATER AND STEAM

During the last few years, it has been shown that the composition of high temperature water or steam systems is affected by the temperature, pressure and rock composition in the hot aquifer. The composition of hydrothermal fluids and gases cover a wide range, but have common characteristics. Water temperatures can be as high as 350°C, and salinity, as indicated by chloride content, may vary from about 30 ppm to 150,000 ppm [Ellis, 1970]. Some components, such as the silica concentration and the potassium/sodium ratio, can be used to estimate the temperature of underground reservoirs.

However, the chemistry of the associated gases is more relevant for this symposium. Most of the explored hydrothermal fields contain a single phase hot water system, the temperature initially increasing with depth near the pressure boiling point. A few systems are vapor dominated. Some data from $Ellis$ [1970] are summarized in Table 1. This information includes only drilled geothermal fields. Larderello in Italy, and the Geysers in California are both vapor dominated; drillholes tap dry steam. The other fields produce a two-phase mixture of steam and water as the original confining pressure is reduced, and only the separated dry steam is used for power production. In the original high temperature reservoir, all the constituents are kept in solution under the confining lithostatic-hydrostatic pressure, though in a few systems there is evidence of phase separation.

TABLE 1.   Composition of Steam from Drillholes [$Ellis$, 1970]

Location	Depth m	Temp. °C	Gas/water mole %	Gas composition* mole %				
				CO_2	H_2S	$CH_4^{\dagger}$	H_2	$N_2^{\ddagger}$
Larderello (average hole)	500	220	2.0	92.8	2.5	$2.0^{\S}$	-	0.55
The Geysers (average hole)	250	200	0.3	69.3	3.0	11.8	12.7	1.6
Wairakei (average hole)	650	260	0.02	91.7	4.4	0.9	0.8	1.5
Broadlands (Hole 11)	760	260	0.2	94.8	2.1	1.2	0.2	1.5
Ngawha (Hole 1)	585	228	6	93.9	0.7	41	0.5	0.8
Salton Sea (average hole)	1750	325	0.1	90	Mostly H_2S, minor $CH_4 + H_2$			

*Other minor components may also be present.

†Includes higher hydrocarbons.

‡Includes Ar.

§This value represents unresolved mixtures of CH_4 and H_2.

The amounts of gas in the water are seen to be variable, but in all cases carbon dioxide is the major component. Some analyses are incomplete (i.e., hydrogen and methane from Larderello have not been separated) and this is a common feature of the literature on thermal areas. Some wells also have measurable amounts of other minor components which are usually omitted from analysis. There is also some variability within fields such as Wairakei, New Zealand, and only average values are quoted here. Ngawha, also in New Zealand, has an unusually high gas content (Table 1). Hot springs (below or near 100°C) can also have a high gas/water ratio, especially if steam separation has occurred and the steam condenses as it heats a perched aquifer. This can lead to relatively high gas flow, with water output varying with local rainfall.

In general, carbon dioxide constitutes about 90% of the gas discharge, and typically the remainder is hydrogen sulfide, methane, hydrogen, and nitrogen, each about the same order of magnitude. Hydrocarbons other than methane are probably not present in high amounts, but detailed analyses are rarely available. A recent analysis by *Gunter and Musgrave* [*1971*] in Yellowstone National Park is given in Table 2. The ratios of methane to higher hydrocarbons are similar to other natural gases [*Colombo et al. 1969*], though the 12% methane content is unusually high for a geothermal gas. Hydrothermal systems such as those described above are not necessarily associated with active volcanism.

Volcanic emanations themselves are rather different, as the magma is closer to the surface and gas temperatures can range from 400°C up to the molten rock temperature [*White and Waring, 1963*]. Gases uncontaminated by groundwater have been shown to have more carbon dioxide than steam, and gas compositions can vary rapidly with time [*Tazieff, 1970*].

Some of the reactions that can occur are shown schematically in Figure 1. In particular, the reactions involving carbon are relatively few. The carbon dioxide may originate from magma, diagenesis of carbonate, organic matter, or dissolved CO_2 in meteoric water. In fact, $\delta^{13}C_{PDB}$ values of thermal CO_2 are in the range +2 to -7°/oo, which is the range of most sedimentary calcite.

At high temperatures, sulfur dioxide is present in greater amounts than hydrogen sulfide, which is the stable form at low temperatures [*Ellis, 1957*]. Hydrogen partial pressure may be controlled by the iron mineral assemblage (W.F. Giggenbach, personal communication). Hydrogen is thus available to reduce carbon dioxide to methane. However, all these reactions are reversible, and the rates of reaction are controlled by temperature. At high temperatures, hydrogen is an important component, and CO, HCl and HF may also occur in the exhaled vapors. High temperature vents, as on White Island, New Zealand, precipitate sulfur at the fumarole mouth, but at red heat, sulfur and also hydrogen may burn on exposure to air.

TABLE 2. Hydrocarbon Content of Y-90, Yellowstone
National Park [*Gunter and Musgrave, 1971*]

	Mole %	Mole % x 10^6	Mole Ratio
CO_2	84.1		
CH_4	12.28		100.
C_2H_6		450.	3.6
C_3H_8		51.	0.42
$n-C_4H_{10}$		25.	0.2
$i-C_4H_{10}$		7.8	0.06
C_2H_2		13.0	0.10
C_2H_4		4.2	0.04

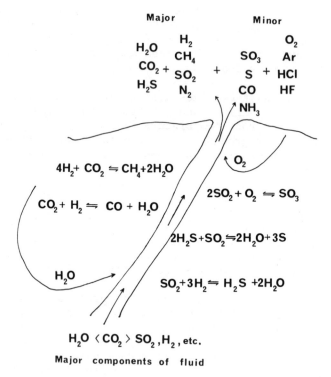

Fig. 1. Some chemical reactions in a fumarolic conduit. Fluids
originate from a parent magma or rocks undergoing metamorphism.

ISOTOPIC CHEMISTRY

Stable isotope equilibria and reaction rates are also temperature dependent. In general, isotopic fractionation between any two molecules decreases with increasing temperature.

Sulfur isotope fractionation between dissolved hydrogen sulfide and the bisulfate ion appears to reach equilibrium only after a few weeks at 200°C or a few days at 300°C at low pH [*Robinson, 1973*], but in low temperature sediments isotopic equilibrium between hydrogen sulfide and oxidized sulfur compounds has not been demonstrated.

Under equilibrium conditions, via the reaction $CO_2 + 4H_2 \rightleftharpoons CH_4 + 2H_2O$, methane formed by reduction of CO_2 with hydrogen is always depleted in ^{13}C relative to carbon dioxide [*Craig, 1953; Bottinga, 1969*], and the fractionation factor decreases with increasing temperature. A number of samples of carbon dioxide and methane have been measured from drillholes at Broadlands and Wairakei, New Zealand (Table 3). Temperatures listed under the subheading of SiO_2 in Table 3 are estimated by dissolved silica concentration [*Mahon, 1966*] and are within 20°C of the downhole temperature, whereas $\delta^{13}C$ temperatures (Table 3) are based on the equilibrium calculations of *Bottinga* [*1969*]. The $\delta^{13}C$ calculated temperatures are higher than the downhole temperatures by as much as 150°C; in hotter wells higher isotope temperatures are generally observed. If the earlier calculations of *Craig* [*1953*] are used, $\delta^{13}C$ temperatures are about 60°C lower.

TABLE 3. Drillhole $^{13}C/^{12}C$ Ratios and Calculated Temperatures from New Zealand Geothermal Areas

| Sample No. | Location | $\delta^{13}C_{PDB}(°/\circ\circ)$ | | Temperature °C | |
		CH_4	CO_2	SiO_2	$\delta^{13}C$
R2597	Wairakei 44	-26.3	-5.7	240	360
R2599	Broadlands 2	-26.2	-7.7	255	400
R2601	Broadlands 7	-25.6	-8.1	265	425
R2791	Broadlands 8	-27.5	-8.2	245	385
R2603	Broadlands 11	-26.1	-7.8	240	405

We have analyzed hot spring and fumarole gases to estimate underground temperatures in areas not yet drilled. Table 4 shows some of these analyses, where the $\delta^{13}C$ temperatures have been estimated from *Bottinga's* [*1969*] calculations. White Island and Mt Tongariro are active volcanoes, with boiling pools and fumaroles. Rotorua and Tikitere hydrothermal systems have shallow wells for use in private bathing pools, and temperature estimates are similar to those of Wairakei. The Maketu spring methane is isotopically very light, with very little carbon dioxide, no detectable hydrogen in its gases, and the spring temperature is 49°C. This is geothermally heated water, but the gas does not seem to be of geothermal origin. It occurs in a swamp at the mouth of a river, and appears to be typical marsh gas and similar to the natural gas from marine sediments.

TABLE 4. Hot Spring δC^{13} ($CH_4 - CO_2$) and Calculated Temperatures Compared to Surface Measured Temperatures

Sample No.	Location	$\delta^{13}C_{PDB}$ (°/oo)		Temperature °C	
		CH_4	CO_2	Surface	$\delta^{13}C$
R507/2	*White Island* Big Donald fumarole	-16.1	-3.1	450	560
R2609	*Mt Tongariro* Ketetahi Springs	-24.7	-8.1	98	450
R2785	*Tikitere* Hell's Gate pool	-28.8	-6.8	77	335
R2788	Parengarenga Spring	-31.1	-8.4	30	325
R2793	*Rotorua* Ngararatuatara pool	-26.3	-7.6	99	400
R2794	Sulphur Point	-35.9	-5.6	88	220
R2784	*Bay of Plenty* Maketu Spring	-70.4	-7.1	49	30

Oxygen and hydrogen also display isotopic fractionation in the reaction $CH_4 + 2H_2O \rightleftharpoons CO_2 + 4H_2$. The $^{18}O/^{16}O$ exchange between carbon dioxide and water is rapid, and re-equilibration readily occurs after sampling. D/H ratios for natural hydrogen and methane gases have rarely been reported in the literature. Table 5 lists some isotope data from the Yellowstone National Park, U.S.A., [*Gunter and Musgrave, 1971*] and from Broadlands, New Zealand. Table 6 compares the temperature estimates for various molecule pairs by use of the data of *Bottinga* [*1969*], which assume only vapor phase reactions. The $\delta^{13}C(CO_2-CH_4)$ temperatures are higher than the $\delta D(CH_4-H_2)$ values, which in turn are higher than the $\delta D(H_2-H_2O)$ values, and best correspond with the measured temperatures. The $\delta D(CH_4-H_2O)$ equilibrium fractionation has a maximum at 350°C, but fractionation values obtained are outside the allowable range. The experimental values may agree better with calculations for liquid water when these are made. Other unpublished data on Kenya geothermal gas samples show similar trends.

These data imply that either the reaction is not in equilibrium at all, or more likely that the isotopic fractionation rates change

TABLE 5. Measured $\delta^{13}C$ and δD Values from Gases at Yellowstone National Park, U.S.A. and Broadlands, New Zealand

	U.S.A. (Y-90 hot spring)		N.Z. (Br-8 drillhole)	
	$\delta^{13}C_{PDB}(°/°°)$	$\delta D_{SMOW}(°/°°)$	$\delta^{13}C_{PDB}(°/°°)$	$\delta D_{SMOW}(°/°°)$
CO_2	-3.8		-8.2	
CH_4	-23.5	-239	-27.5	-179
H_2		-661		-457
H_2O		-140		-40

TABLE 6. Temperatures Derived from Isotopic Equilibrium Calculations Using the Thermodynamic Data of *Bottinga* [*1969*]

	$\delta^{13}C(CO_2-CH_4)$	$D(CH_4-H_2)$	$\delta D(CH_4-H_2O)$	$\delta D(H_2-H_2O)$	Measured
Y-90	380	125	-	105	81
Br-8	385	325	-	265	245

as the gas approaches the surface and cools. This would suggest that the rates of reaction for the approach to isotopic equilibrium decrease in the order $\delta D(H_2-H_2O)$, $\delta D(H_2-CH_4)$, $\delta^{13}C(CO_2-CH_4)$. The $\delta^{13}C$ temperature would then be expected to best represent the main high temperature gas reservoir.

RELEVANCE TO MARINE GASES

The above discussion indicates that geothermal gas compositions are considerably different from natural gases produced by microbial action in marine sediments. It is possible, however, that gas mixtures of different origins may occur. Gas bubbles rising from the sea bottom have been reported by *McCartney and Bary* [1965] from the Saanich Inlet, where the gas is probably methane of microbial origin formed in the sediments [*Nissenbaum et al. 1972*]. Bubbles detected visually and by echo-sounder are rising from the floor of the Bay of Plenty, New Zealand, and these have been interpreted as geothermal gas emanations [*Duncan and Pantin, 1969; Glasby, 1971*]. Due to their sporadic distribution, none of these gases has yet been collected.

Geothermal fluids passing through marine sediments will react with silicates, carbonates and organic matter, and they will also be absorbed by solution in interstitial water. Further solution and oxidation of acid gases will occur as gases rise through the sea water column so that only methane, hydrogen, nitrogen and rare gases can be expected to reach the sea surface. These effects have been observed in the fresh water Lake Rotoiti in the New Zealand thermal area, where bubbles were recently noted and gas was collected. Approximate analyses showed the H_2/CH_4 ratio near to unity, and the $\delta^{13}C$ of the methane as $-28°/_{oo}$, which are similar to values obtained for gases from the nearby Tikitere thermal area. Undoubtedly this gas is of geothermal origin, as a thermal gradient of 50°C m^{-1} was measured in the sediments under 90 meters of water (Ian Calhaem, personal communication).

Geothermal heating is likely to occur under marine sediments such as in the Gulf of California, the Red Sea and mid-ocean ridges. It may also occur in regions where natural gas has formed, and will result in the production of mixtures of thermal and biogenic gases. Such processes are likely to occur at plate margins, where subduction of sediments containing biogenic gas might result in production of complex mixtures.

ACKNOWLEDGMENTS

I am especially grateful to Mrs. M.A. Cox for technical assistance in the laboratory, and to Dr. W.F. Giggenbach, Chemistry

Division, D.S.I.R., for critical comments on this paper. Other colleagues at the Institute of Nuclear Sciences also made valuable contributions.

Contribution 604 of the Institute of Nuclear Sciences.

REFERENCES

Bottinga, Y., Calculated fractionation factors for carbon and hydrogen isotope exchange in the system calcite-carbon dioxide-graphite-methane-hydrogen-water vapour, *Geochim. et Cosmochim. Acta, 33,* 49, 1969.

Colombo, U., F. Gazzarrini, R. Gonfiantini, E. Tongiorgi, and L. Caflisch, Carbon isotope study of hydrocarbons in Italian natural gases, in *Advances in Organic Geochemistry,* vol. 4, edited by P. A. Schenck, p. 499, Pergamon Press, New York, 1969.

Craig, H., Geochemistry of carbon isotopes, *Geochim. et Cosmochim. Acta, 3,* 53, 1953.

Duncan, A. R., and H. M. Pantin, Evidence for submarine geothermal activity in the Bay of Plenty, *N. Z. J. Mar. Freshwater Res., 3,* 602, 1969.

Ellis, A. J., Chemical equilibrium in magmatic gases, *Amer. J. Sci., 255,* 416, 1957.

Ellis, A. J., Quantitative interpretation of chemical characteristics of hydrothermal systems, *Geothermics Spec. Issue 2,* p. 516, 1970.

Glasby, G. P., Direct observations of columnar scattering associated with geothermal gas bubbling in the Bay of Plenty, New Zealand, *N. Z. J. Mar. Freshwater Res., 5,* 483, 1971.

Gunter, B. D., and B. C. Musgrave, New evidence on the origin of methane in hydrothermal gases, *Geochim. et Cosmochim. Acta, 35,* 113, 1971.

McCartney, B. S., and B. M. Bary, Echo-sounding on probable gas bubbles from the bottom of Saanich Inlet, British Columbia, *Deep-Sea Res., 12,* 285, 1965.

Mahon, W. A. J., Silica in hot water discharged from drillholes at Wairakei, New Zealand, *N. Z. J. Sci., 9,* 135, 1966.

Nissenbaum, A., B. J. Presley, and I. R. Kaplan, Early diagenesis in a reducing fjord, Saanich Inlet, British Columbia - I. Chemical and isotopic changes in major components of interstitial water, *Geochim. et Cosmochim. Acta, 36,* 1007, 1972.

Robinson, B. W., Sulphur isotope equilibrium during sulphur hydrolysis at high temperatures, *Earth Planet. Lett.*, *18*, 443, 1973.

Tazieff, H., New investigations on eruptive gases, *Bull. Volcanol.* *34*, 421, 1970.

White, D. E., and G. A. Waring, Volcanic emanations, in *Data of Geochemistry*, edited by M. Fleischer, Chap. K, Geol. Survey Professional Paper 440-F, U. S. Government Printing Office, Washington, 1963.

THE NATURE AND OCCURRENCE OF CLATHRATE HYDRATES

Stanley L. Miller

Department of Chemistry
University of California at San Diego
La Jolla, California 92037

ABSTRACT

Clathrate hydrates are crystalline compounds in which an expanded ice lattice forms cages that contain gas molecules. There are two principal gas hydrate structures. Structure I, with a 12 Å cubic unit cell, contains 46 water molecules and 8 cages of two types, giving an ideal formula (for CH_4) of $CH_4 \cdot 5\text{-}3/4H_2O$. The actual formula contains somewhat more water as the cages are not completely filled. Examples of gases that form Structure I hydrates are ethane, N_2, O_2, Ar, Xe, CH_3Cl, H_2S. Structure II, with a 17 Å cubic unit cell, contains 136 water molecules, and 8 large cages and 16 small cages. This gives an ideal formula of, for example, $CHCl_3 \cdot 17H_2O$. Other molecules that form a Structure II hydrate include propane, ethyl chloride, acetone, and tetrahydrofuran. The conditions of pressure and temperature for hydrate formation are discussed. The statistical-mechanical treatment of hydrate stabilities shows that the cages are not completely occupied; thus the clathrate hydrates are non-stoichiometric compounds.

Methane hydrate is likely to be a major constituent of the planets Uranus and Neptune. The hydrates of ammonia ($2NH_3 \cdot H_2O$, $NH_3 \cdot H_2O$, and $NH_3 \cdot 2H_2O$), which are stoichiometric compounds, are also likely to occur on these planets. Comets are likely to contain similar hydrates. The ice cap of Mars, which is mostly solid CO_2, should contain CO_2 hydrate. On the earth, a hydrate of air $[(N_2, O_2) \cdot 6H_2O]$ occurs in the Antarctic ice cap. Methane hydrate has been reported to be present in some oceanic sediments based on sound-velocity data.

The conditions for the formation of methane hydrate from pure

water and methane gas are given, and the hydrate forming conditions
are calculated when the hydrostatic pressure is greater than the
dissociation pressure of the hydrate. The amount of methane dissolved in the water is calculated under these conditions, and it is
clear that bubbles of methane gas need not be present in the water
for the hydrate to be stable. These figures can be applied to the
stability of methane hydrate in oceanic sediments by correcting
for the presence of the dissolved salts. The effect of other hydrate forming gases (i.e., H_2S) on the stability of methane hydrate
is also calculated.

INTRODUCTION

A clathrate compound is one in which a crystal lattice contains
cages (or voids) that can incorporate guest molecules (i.e., CH_4).
The clathrate hydrates, sometimes referred to as gas hydrates, are
a special case of clathrate compounds in which the framework consists of water molecules [for reviews see *van der Waals and Platteeuw, 1959; Jeffrey and McMullan, 1967; and Davidson, 1973*]. The
clathrate hydrates can be considered as low pressure forms of ice,
which are only stable when a gas molecule is present in the cages.

Molecules such as urea, thiourea, phenol, hydroquinone (p-
dihydroxybenzene, sometimes called quinol) form crystal lattices
with cages in them that can incorporate guest molecules [for reviews
on clathrates see *Mandelcorn, 1959; Hagen, 1962; Powell, 1964; and
Bhatnagar, 1968*]. Silica can form clathrate structures, not only
in the case of the zeolites (molecular sieves are an example of this),
but also a structure analogous to one of the clathrate hydrates
[*Kamb, 1965a*]. This discussion will be confined largely to the clathrate hydrates, with particular emphasis on the gas hydrates.

The first clathrate hydrate was discovered by Sir Humphry Davy
in 1810, who cooled an aqueous solution saturated with chlorine gas
below 9°C [*Davy, 1811*]. He observed that crystals of an ice-like
material formed that contained about one Cl_2 for each ten water
molecules. The field lay largely dormant until the period 1880-1910
when Villard and de Forcrand, as well as others, investigated a
large number of gases that formed hydrates (i.e., CH_4, CO_2, C_3H_8,
etc). The field again remained largely dormant until 1935-1945 when
the clathrates of natural gas were found to precipitate in natural
gas pipelines, thereby clogging them [*Hammerschmidt, 1934; Deaton
and Frost, 1946*]. This problem was solved by the simple expedient
of drying the natural gas before putting it into the pipeline.

The modern investigations of clathrates begin with the work of
Stackelberg and co-workers [*1949-1954*], who showed by X-ray dif-
fraction that there were two types of gas hydrates (Structure I and
Structure II). Stackelberg showed that the gas hydrates were

clathrate compounds, the nature of which had been demonstrated by
Palin and Powell [*1945; 1948a,b*] for the hydroquinone clathrates.

Some progress was made by Stackelberg in determining the crystal structure of the hydrates, but the first determination of the
Structure I hydrate was done by *Pauling and Marsh* [*1952*]. The correct structure for the Structure II hydrates was guessed by model
building [*Claussen, 1951a,b*] and subsequently determined by X-ray
diffraction [*Stackelberg and Müller, 1951; Mak and McMullan, 1965*].

Calculations of clathrate hydrate properties, based on statistical mechanical treatment by *Barrer and Stuart* [*1957*] were made by
van der Waals and Platteeuw [*1959*].

In 1960, Jeffrey and co-workers began a program of determining
the structure of a large number of amine hydrates [*Jeffrey and Mc-
Mullan, 1967*], including uncharged amines (i.e., t-butyl amine) and
cationic amines (i.e., $(C_5H_{11})_4N^+F^-$). A wide variety of hydrate
structures, in addition to the Structure I and Structure II hydrates,
was determined. This discussion will not consider these neutral
amine and cationic amine hydrates.

Recent work on clathrate hydrates include some very accurate
measurements of dissociation pressures [*Glew, 1960*], their use in
desalination of sea water [*Parker, 1942; Donath, 1960; Williams,
1961; Glew, 1962a; Hess and Jones, 1964; Glew, 1967; and Barduhn,
1967*], discussions of their occurrence in the solar system [*Miller,
1961a, 1969*], numerous studies of dielectric relaxation and nuclear magnetic resonance [*Davidson, 1973*], and the use of gas hydrates as a basis for a theory of general anesthesia [*Pauling, 1961;
Miller, 1961b*].

Clathrate hydrates have been used to stabilize free radicals
at low temperature and to study their electron paramagnetic resonance
[*Goldberg, 1963*]. The hydrates are frequently used as model compounds in discussions of non-polar gases dissolved in liquid water
[*Frank and Evans, 1945; Frank, 1970; and Claussen and Polglase,
1952*], and also in discussions of the hydrophobic bond that is responsible in part for the tertiary structure of proteins [*Kauzmann,
1959; Klotz, 1958; Klotz, 1960; and Kavanau, 1964*].

Water alone can form two different self-clathrate structures.
Ice VII consists of two interpenetrating cubic ice lattices [*Kamb
and Davis, 1964*]. There are no hydrogen bonds between the two different lattices that are held in place by van der Waals forces.
Ice VI also consists of two interpenetrating lattices that are not
hydrogen bonded to each other [*Kamb, 1965b*].

HYDRATE STRUCTURES

Structure I Hydrate

Gases that form this hydrate include Ar, Kr, Xe, CH_4, C_2H_6, C_2H_4, CH_3Cl, Cl_2 and cyclopropane. Molecules such as $CHCl_3$, C_2H_5Cl and propane are too large to fit in the cavities of the Structure I hydrate and so form Structure II hydrates.

The framework of the Structure I hydrate is shown in Figure 1. There are 46 molecules in the unit cell that form 2 pentagonal dodecahedra (12 sided polyhedron) and 6 tetrakaidecahedra (14 sided polyhedron). Molecules 5.1 Å or less in diameter will fit into the pentagonal dodecahedra (CO_2 and CH_4, but not CH_3Br), and molecules 5.8 Å or less in diameter will fit into the tetrakaidecahedra (CH_4 and CH_3CH_3, but not $CH_3CH_2CH_3$). The ideal formula for a hydrate with both types of cavities filled is 46/8 = 5-3/4 (i.e., $CH_4 \cdot 5\text{-}3/4H_2O$). If only the tetrakaidecahedra are occupied, the ideal formula is 46/6 = 7-2/3 (i.e., cyclopropane$\cdot 7\text{-}2/3H_2O$).

The actual formulas differ from the ideal formulas because the cavities are not 100% occupied by gas molecules. For example, direct analysis [*Frost and Deaton, 1946*] gives a formula of $CH_4 \cdot 7.1H_2O$ instead of the ideal $CH_4 \cdot 5\text{-}3/4H_2O$; however, the hydrate formula derived from a thermodynamic treatment of the dissociation pressures gives a formula close to the ideal formula [*Glew, 1962b*]. The determination of the hydrate formulas is a difficult task and there is considerable uncertainty as to correct formulas. It is clear from the statistical mechanical treatment of hydrate stabilities that there is never 100% occupancy of the cavities. However, it appears that for most hydrates the actual formula is not far from the ideal formula.

It should be noted that the water molecules forming the cavities do not interact specifically with the encaged molecule. In other words, the structure of methane hydrate is the same as carbon dioxide hydrate. Thus, mixed hydrates of methane and carbon dioxide can be prepared by using a mixture of gaseous methane and carbon dioxide. The ideal hydrate formula is then $(CH_4, CO_2) \cdot 5\text{-}3/4H_2O$, rather than a mixture of $CH_4 \cdot 5\text{-}3/4H_2O$ and $CO_2 \cdot 5\text{-}3/4H_2O$. A mixed hydrate is thus a solid solution of gases in the hydrate framework.

Structure II Hydrate

This structure contains 136 molecules of water in the unit cell of 17 Å. There are 16 pentagonal dodecahedra very similar to those of the Structure I hydrate and 8 hexakaidecahedra (16 sided polyhedron). Molecules with diameters of 6.7 Å or less will fit into the

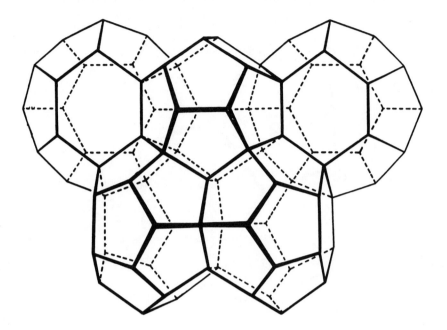

Fig. 1. The 12 Å hydrate lattice. The pentagonal dodecahedron (upper center) is formed by 20 water molecules. The tetrakaidecahedra formed by 24 water molecules have 2 opposite hexagonal faces and 12 pentagonal faces.

hexakaidecahedra ($CH_3CH_2CH_3$, but not $CH_3CH_2CH_2CH_3$). The unit cell cannot be visualized with less than a three dimensional model, but the hexakaidecahedron is shown in Figure 2. The ideal formula for a Structure II hydrate is 136/8 = 17 (i.e., $CHCl_3 \cdot 17H_2O$). The actual formula appears to be close to the ideal formula [$Glew$, 1960]. Chloroform is much too large to fit into the pentagonal dodecahedra, but if a smaller molecule such as H_2S is mixed with the chloroform, a mixed hydrate, usually termed a double hydrate, is formed (i.e., $CHCl_3 \cdot 2H_2S \cdot 17H_2O$). In a double hydrate, the large molecule ($CHCl_3$) is confined to the large cavity and the small molecule (H_2S) is largely, but not entirely, confined to the small cavities. This differs from a mixed hydrate, i.e., [$(CH_4, CO_2) \cdot 5\text{-}3/4H_2O$], where the methane and carbon dioxide are present in both the large and small cavities. The mole fraction of carbon dioxide is expected to be larger in the large cavities than in the small cavities because carbon dioxide is larger than methane. Double hydrates are only known for Structure II hydrates, although a Structure I double hydrate is a possibility (i.e., cyclopropane $\cdot 1/3H_2S \cdot 7\text{-}2/3H_2O$).

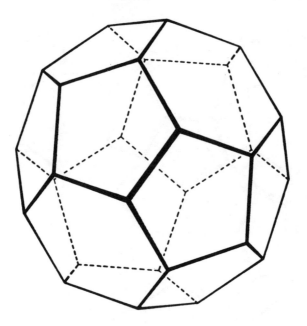

Fig. 2.   The hexakaidecahedron formed by 28 water molecules in the
17 Å lattice.   There are 4 hexagonal faces and 12 pentagonal faces.

DISSOCIATION PRESSURES AND PHASE DIAGRAMS

The dissociation pressures of gas in a hydrate, which can be
considered as the vapor pressure of the hydrate, is a definite value
when the hydrate is in equilibrium with ice and the gas (or with
water and the gas).   The dissociation pressure of a number of hydrates at 0°C is given in Table 1.   Just as the vapor pressure of a
pure substance varies with temperature, that of a hydrate does also,
but in a somewhat more complicated way.   Figure 3 shows the phase
diagram of methane hydrate.   At -20°C, the phase diagram shows that
for pressures less than 13.5 atm of CH_4,   ice + $CH_4(g)$ are the
only stable phases.   If the pressure is raised to 13.5 atm,   ice +
CH_4 hydrate + $CH_4(g)$ are in equilibrium.   If the CH_4 pressure is
raised above 13.5 atm, all the ice will be converted to hydrate.
Above 0°C, the equilibria involve liquid water.   For example, at
+16°C, the dissociation pressure is 136 atm.   Below this pressure
CH_4 hydrate is unstable; if the pressure of CH_4 is raised above
136 atm, all the water will be converted to methane hydrate.   The
quadruple point, where   ice + CH_4 hydrate + water + $CH_4(g)$ are in
equilibrium, is slightly below 0°C (-0.19°C) and 26 atm.   Because
the critical point of CH_4 is at -82.5°C, the pressure (and fugacity)
of CH_4 can be increased indefinitely without forming liquid CH_4;

TABLE 1. Examples of Gas Hydrate Dissociation Pressures

Gas	P_{diss}^{atm} $(0°C)$	Hydrate Structure
N_2	160	I
O_2	120	I
A	95.5	I
Kr	14.5	I
Xe	1.50	I
CF_4	41.5	I
CH_4	26.0	I
C_2H_6	5.2	I
$CH_2-CH_2-CH_2$	0.70	I (metastable)
CO_2	12.5	I
H_2S	0.92	I
Cl_2	0.32	I
Br_2	0.056	*
C_3H_8	1.74	II
$CH_2-CH_2-CH_2$	0.62	II
$CHCl_3$	0.065	II
C_2H_5Cl	0.26	II
$CBrClF_2$	0.19	II

*Bromine apparently forms a tetragonal hydrate structure containing pentagonal dodecahedra, pentakaidecahedra and hexakaidecahedra [*Jeffrey and McMullan*].

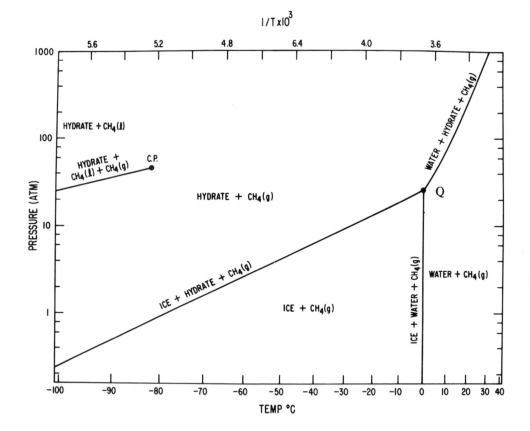

Fig. 3.   The phase diagram of methane hydrate.   The line   hydrate
+ CH_4(1) + CH_4(g)   ends very close to the critical point of methane
(191°K).   Q   is the quadruple point   (ice + hydrate + water +
CH_4(g)).

thus there is no limit on the temperature at which this hydrate can
be prepared (at least until ice VI and VII become stable relative
to liquid water).   Methane hydrate has been prepared at temperatures
as high as 47°C (3900 atm) [*Marshall et al. 1964*].   Because the
slope of the line describing the vapor pressure of liquid and solid
CH_4 (very close to the line   hydrate + CH_4(1) + CH_4(g))   is less
than the slope of the dissociation pressure curve (Figure 3), methane hydrate is stable to 0°K.

Figure 4 shows a somewhat different case of the CO_2 hydrate.
In this case the critical point is at 31.0°C, so that the dissociation pressure curve meets the vapor pressure curve at 10.20°C and

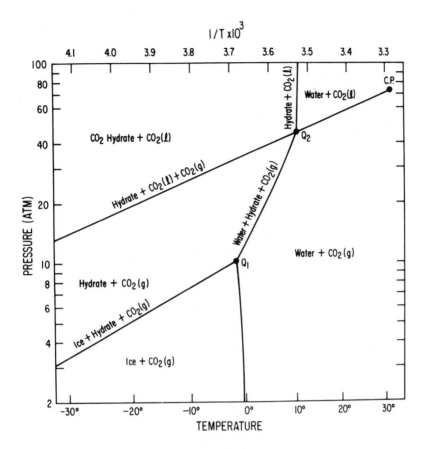

Fig. 4. The phase diagram of carbon dioxide hydrate near 0°. C.P. is the critical point of carbon dioxide (31.0°C and 72.8 atm). Q_1 is the quadruple point [ice + water + hydrate + CO_2(g)] at -1.77°C and 10.20 atm, Q_2 is the quadruple point [water + hydrate + CO_2(1) + CO_2(g)] at 10.20°C and 44.50 atm.

44.50 atm (Q_2). The hydrate is not stable above this temperature, except at extremely high pressures with the equilibrium CO_2 hydrate + water + CO_2(liq). Thus, CO_2 hydrate has been prepared at 19.5°C and 1840 atm [*Takenouchi and Kennedy, 1965*].

Figure 5 shows the low temperature region of the CO_2 hydrate. In this case the slope of the vapor pressure curve of CO_2 (dry ice) is greater than the slope of the hydrate dissociation pressure curve, and thus the two curves meet at approximately 121°K. Below this temperature CO_2 hydrate is unstable relative to ice and solid CO_2 [*Miller and Smythe, 1970*].

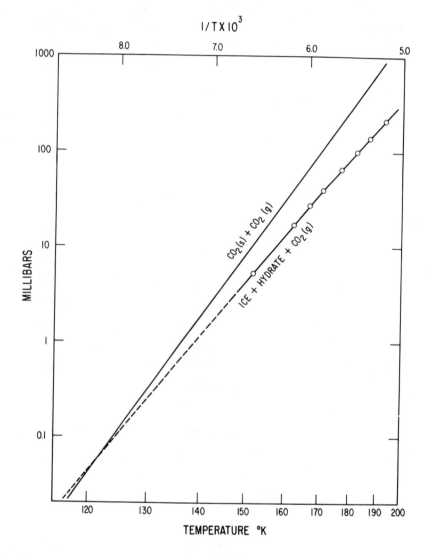

Fig. 5. The dissociation pressure of CO_2 hydrate and the vapor pressure of solid CO_2. The circles are the experimental dissociation pressure measurements. Above about 121°K, CO_2 hydrate is stable at pressures equal to or greater than the dissociation pressure. Below 121°K, CO_2 hydrate is unstable with respect to decomposition to ice and $CO_2(s)$.

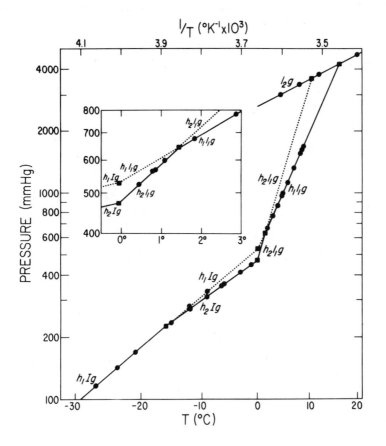

Fig. 6. Dissociation pressure curve for cyclopropane hydrate: ●, selected data points; ■, quadruple points; ——, stable hydrate dissociation pressure curves; ···, metastable hydrate dissociation pressure curves.

The more complicated case of cyclopropane hydrate is shown in Figure 6, where both Structure I and Structure II hydrates are formed [*Hafemann and Miller, 1969a*]. Below -16°C, Structure I hydrate is stable; between -16° and +4° Structure II hydrate is the stable form; above 4° Structure I again becomes stable until 16° where the critical decomposition point is reached. It was possible to measure metastable points of the Structure I hydrate above -16°; these are also shown in Figure 6.

The hydrates of cyclopropane were prepared with D_2O (deuterio-hydrates) [*Hafemann and Miller, 1969b*]. Figure 7 shows this phase diagram which is similar to that of the H_2O case. Below 0°C the

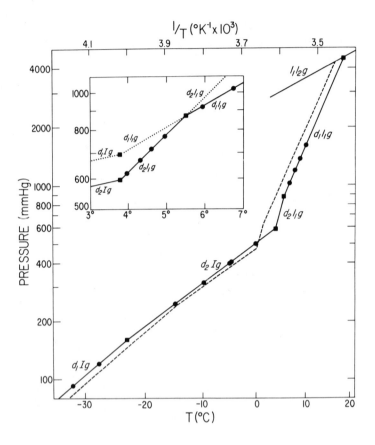

Fig. 7. Dissociation pressure curve for cyclopropane deuteriohydrate: ●, selected data points; ■, quadruple points; —,
stable deuteriohydrate dissociation pressure curve and cyclopropane
vapor pressure curve; ···, metastable deuteriohydrate dissociation
pressure curves; ---, stable cyclopropane hydrate dissociation pressure curve.

deuteriohydrate has a higher dissociation pressure than the hydrate,
but above 4°C the deuteriohydrate dissociation pressure is lower.
One can say that the dissociation pressure with liquid D_2O 'should'
be lower than that for H_2O because D_2O is 'more structured' than
H_2O and therefore this 'extra structure' should make clathrate formation easier. From a thermodynamic standpoint, the lower dissociation pressure with liquid D_2O would be attributed to the higher
melting point of D_2O and the relationships of the slopes of the
ice-hydrate-gas and water-hydrate-gas equilibria rather than to any
structure in the liquids.

STATISTICAL MECHANICS OF CLATHRATE FORMATION

The statistical mechanical treatment of clathrate formation [*van der Waals and Platteeuw, 1959; Barrer and Stuart, 1957; McKoy and Sinanoglu, 1963*], of which only the basic results will be presented here, treats hydrate formation as the sum of two reactions

ice $\rightleftharpoons$ empty hydrate lattice $\qquad$ ΔG = 0.2 kcal per mole H_2O $\qquad$ (1)

empty hydrate lattice + gas $\rightleftharpoons$ hydrate $\qquad$ ΔH = -4 to -8 kcal per mole of gas $\qquad$ (2)

---

ice + gas $\rightleftharpoons$ hydrate $\qquad$ ΔG = 0 at equilibrium
ΔH = -4 to -8 kcal
per mole of gas $\qquad$ (3)

The free energy of formation of the empty hydrate lattice from Ice I_h is unfavorable by ≈ 0.2 kcal per mole of water (1.15 kcal per mole of gas in the ideal Structure I hydrate). This unfavorable reaction together with the unfavorable entropy of compression of the gas is overcome by the enthalpy gained in putting the gas molecule into the cavity. The statistical mechanical treatment uses various methods to calculate the enthalpy (and entropy) of the second reaction, using various potential interactions (i.e., the Lennard-Jones and Devonshire 12,6 potential). When this is evaluated theoretically, and two arbitrary constants are assigned (ΔG for reaction (1) and a normalization factor based on the argon dissociation pressure), the calculation of the dissociation pressure for all Structure I hydrates can be carried out. The theoretical dissociation pressures agree surprisingly well (within about 20%) for the noble gases and spherical molecules (except for CF_4), and within a factor of 5 or better for molecules such as ethane, ethylene and acetylene.

The statistical mechanical treatment shows that the condition for stability of a Structure I hydrate in equilibrium with ice and the gas is

$$\frac{2}{46} \ln (1-y_1) + \frac{6}{46} \ln (1-y_2) = -\Delta G/RT$$

where ΔG is the free energy for reaction (1), y_1 is the fraction of the pentagonal dodecahedra occupied by the gas molecules, and y_2 is the fraction of the tetrakaidecahedra occupied. For the

Structure II hydrates, the 2/46 and 6/46 become 16/136 and 8/136, respectively. This equation shows that the cages can never be completely occupied (unless ΔG were infinite). Thus the clathrate hydrates are non-stoichiometric compounds, although many hydrates have occupancy numbers close to 100%, and so have compositions close to the ideal formulas.

NATURAL OCCURRENCE OF GAS HYDRATES

The occurrence of gas hydrates in natural gas pipelines was mentioned previously. These hydrates are mixed Structure I hydrates of CH_4 and C_2H_6, along with Structure II double hydrates of propane and methane (ideally $C_3H_8 \cdot 2CH_4 \cdot 17H_2O$).

In the solar system, the most abundant occurrence of hydrates must be on the planets Uranus and Neptune [*Miller, 1961a*]. Their densities are 1.56 and 2.22, respectively. These figures indicate, when allowance is made for the compression of material in the interior of the planet, that these planets must be made up largely of CH_4, NH_3 and H_2O. Methane has been detected in the atmospheres of Uranus and Neptune. Ammonia is believed to be frozen out below the cloud layer on these planets. The pressure at the cloud layer of Uranus is about 0.1 atm and the temperature is about 90°K. By extrapolating the dissociation pressure curve of methane hydrate measured at higher temperatures, the dissociation pressure of the hydrate is calculated to be 2×10^{-6} atm at 90°K. Thus, methane hydrate is very stable relative to solid or liquid methane (the triple point of methane is 90.7°K and 0.12 atm). In fact, all the water on the planet should be converted to methane hydrate if there is an excess of methane, or all the methane should be converted to hydrate if there is an excess of water (the relative proportions of CH_4, H_2O and NH_3 are not known accurately).

This picture is complicated by the reaction of NH_3 with water to form the ammonia hydrates, the phase diagram for which is shown in Figure 8. The ammonia hydrates, $2NH_3 \cdot H_2O$, $NH_3 \cdot H_2O$ and $NH_3 \cdot 2H_2O$, are stoichiometric compounds rather than clathrates. The crystal structures of $2NH_3 \cdot H_2O$ and $NH_3 \cdot H_2O$ show that these are hydrogen bonded solids [*Siemons and Templeton, 1954; Olovsson and Templeton, 1959*]. Except for their thermal properties [*Hildebrand and Giauque, 1953; Chan and Giauque, 1964*], which show ordered hydrogen bonds at 0°K, these 'ices' have received little attention, and they may prove to be very interesting compounds in their own right. Even with this complication, it seems very likely that Uranus and Neptune must be composed of methane hydrate, the ammonia hydrates, and perhaps water ice.

Jupiter and Saturn are believed to be composed largely of hydrogen and helium, but methane and ammonia have been detected in

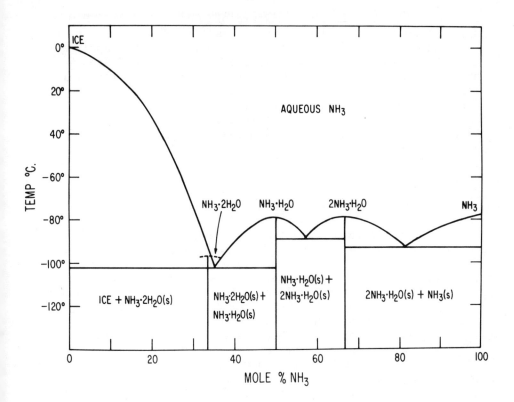

Fig. 8. The phase diagram of NH_3-H_2O system at low temperature. The melting points of NH_3, $2NH_3 \cdot H_2O$ and $NH_3 \cdot H_2O$ are $-77.80°$, $-78.83°$ and $-79.00°$, respectively. The hydrate $NNH_3 \cdot 2H_2O$ melts (incongruently) at $-97.06°$. Its composition-melting curve is shown in the figure as the dotted line.

their atmospheres. It seems possible that methane hydrate is present on these planets along with the ammonia hydrates [*Miller, 1961a; Lewis, 1969*], but not enough is known about the structure of their atmospheres to make an accurate prediction.

Mercury has no atmosphere and almost certainly no water. There is a great deal of CO_2 on Venus and small amounts of water, and thus the formation of a CO_2 hydrate is possible, but the temperature and pressure conditions as presently accepted are not favorable for this.

Comets seem another likely place for the occurrence of hydrates [*Miller, 1961a; Delsemme and Swings, 1952; Delsemme and Wegner, 1970; and Delsemme and Miller, 1970*]. The most popular model of comets is that of a mixture of 'ices' (a mixture of CH_4, NH_3 and H_2O ices), along with some silicate particles [*Whipple, 1951*]. Because considerations discussed under Uranus and Neptune would apply here, it is expected that methane hydrate and the ammonia hydrates would be present. Little is known about the chemical compositon of interstellar dust, but on the basis of the cosmic abundances of the elements, it would be expected to contain solid methane, ammonia, and water. The temperature of the interstellar dust is usually given as 3°K, and thus methane hydrate and the ammonia hydrates are thermodynamically stable. However, the situation here is complicated by the kinetics of hydrate formation and the presence of a strong ultraviolet radiation field.

It appears very likely that the CO_2 hydrate is present in the ice cap of Mars [*Miller and Smythe, 1970*]. The Mariner 6 and 7 missions in 1967 first reported a temperature of the ice cap of 153°K, and a partial pressure of CO_2 of 6.5 mbars [*Neugebauer et al. 1969; Kliore et al. 1969*]. The vapor pressure of dry ice is 13.1 mbars at this temperature. In order for the ice cap to be dry ice at 6.5 mbars, the temperature would have to be 148°K. The 153°K and 6.5 mbars of CO_2 can be accounted for on the basis of a CO_2 hydrate in equilibrium with ice [*Miller and Smythe, 1970*]. Figure 5 shows the phase diagram of CO_2 hydrate in this temperature region. The temperature data for the ice cap were subsequently refined and the temperature revised downward to 148°K and 6.5 mbars, which is in accord with the vapor pressure of dry ice [*Neugebauer et al. 1971*]. This situation still leaves the CO_2 hydrate stable, but the phases in equilibrium are CO_2 hydrate + CO_2(solid) + CO_2 (gas) rather than CO_2 hydrate + ice + CO_2(gas). It is not certain that equilibrium would be attained as the formation of CO_2 hydrate at 148°K is very slow unless the ice is finely divided [*Miller and Smythe, 1970*]. However, it would be expected that ice precipitating from the Martian atmosphere would be very finely divided.

The first reported natural occurrence of clathrate hydrates on the earth (aside from the natural gas pipelines) is a hydrate of air in an antarctic ice core [*Miller, 1969*], which was drilled 2164 meters to bedrock [*Gow et al. 1968*]. It was observed that bubbles of air were present in the ice core near the top, and they became smaller as the depth increased by the compression of the bubble due to the overlying weight of ice. However, the bubbles became smaller than predicted at 900 meters by the compression, and by 1200 meters the bubbles entirely disappeared and remained absent to the bottom of the ice core.

The dissociation pressures of nitrogen, oxygen, and air hydrates are shown in Figure 9, together with the temperature and pressures in the ice core. The data show that the air hydrate should begin to form at 800 meters depth and the bubbles should completely disappear at 850 meters. Thus there is a discrepancy of about 200 meters.

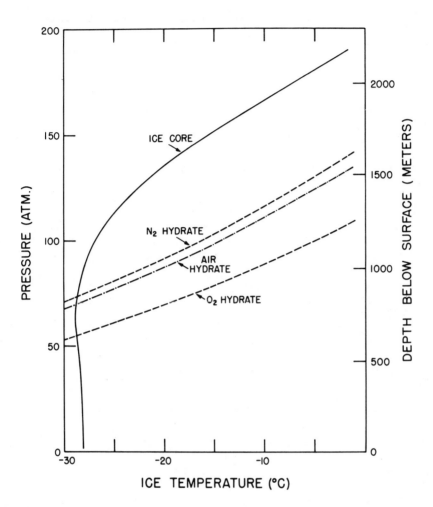

Fig. 9. The relation between temperature, pressure and depth in the Antarctic ice core. The dissociation pressures of N_2, O_2 and air hydrates are also shown.

Although the discrepancy between the predicted depth for the disappearance of bubbles and the observed depth is not great, it might be accounted for by errors in the hydrate data, the ice temperature, density, or depth. The hydrate might also have decomposed to form bubbles in the 800 to 1200 meters part of the core when the pressure was released on the ice because of the way the ice crystals were packed. The discrepancy is more likely to result from the pressure in the bubbles being less than the hydrostatic pressure. When air hydrate forms, the pressure in the bubble decreases. The restoration of the air pressure in the bubble to the hydrostatic pressure will be slow because of the high viscosity of ice [*Nye*, *1953; Kingery, 1963*]. The observation that gas is released only when the deep ice melts implies that the structural strength of the ice crystal mass is greater than the dissociation pressure of the hydrate.

The amount of hydrate can be calculated on the assumption that 10% of the volume of the ice is gas bubbles when the snow is compacted into ice. On the basis of the formula $(N_2, O_2) \cdot 6H_2O$ 0.06% of the ice would be in hydrate form. This amount of cubic crystals of air hydrate would not be easily detected in the presence of the hexagonal ice.

A most exciting occurrence of a natural gas hydrate has been reported by the Russians in Siberia [*Ottawa Citizen, 1971; New York Times, 1972; Makogon et al. 1971*]. The reports are fragmentary, but it appears that there are fields of natural gas hydrates beneath the permafrost. If such hydrate deposits occur in Siberia, there is a reasonable possibility that such fields exist in Northern Canada and Alaska. In view of the shortage of natural gas, a problem that is likely to get worse in the future, the discovery and exploitation of such natural gas hydrate deposits will have great economic importance as well as chemical and geological interest.

It has recently been proposed that there are hydrates of methane in deep sea sediments [*Stoll et al. 1971*]. The methane hydrates would probably be mixed hydrates of CH_4, CO_2, C_2H_6, H_2S and other gases that are in sea water. The evidence is based only on observations of anomalous sound absorption in certain sediments which could be accounted for by clathrate hydrates. Some sediment cores have large amounts of gas in them, sufficient in many cases to blow the core out of the coring device. The possibility of having hydrates in deep sea sediments is quite reasonable, but evidence more direct than sound absorption is needed to establish this.

THE STABILITY OF METHANE HYDRATE IN OCEANIC SEDIMENTS

The calculation of the conditions for stability of gas hydrates in the ocean is more complicated than for a system containing only water and pure gas. The problem will be treated first for pure methane hydrate.

The dissociation pressure of methane hydrate between 0° and 10°C is given by

$$\log {}_{10}P_{atm} = 1.4156 + 0.04160t + 2.93 \times 10^{-4} t^2$$

where t is in degrees centrigrade. This equation is a least squares fit of the data of *Deaton and Frost* [1946], *Otto and Robinson* [1960] and *Jhaveri and Robinson* [1965]. This equation gives P_{diss} = 26.04 atm at 0°C for the dissociation pressure of methane hydrate. At 26.04 atm, the fugacity/pressure ratio [*Din, 1961*] is f_{diss}/P_{diss} = 0.9413; hence f_{diss} = 24.51 atm.

To calculate the fugacity at any hydrostatic pressure we first calculate the dissociation fugacity at 1 atm pressure, using the method of *Glew* [1960]. For the reaction

$$CH_4 \cdot nH_2O \text{ (hydrate)} = CH_4(g) + nH_2O(liq)$$

the fugacity as a function of hydrostatic pressure and water activity is

$$\ln f_{P \text{ atm}} = \ln f'_{1 \text{ atm}} - n \ln a_w - \frac{(P_{hydro} - 1)}{RT}(nV_{H_2O} - V_h) \tag{1}$$

where P_{hydro} is the hydrostatic pressure, V_{H_2O} is the molar volume of liquid water, V_h is the volume of hydrate that contains 1 mole of CH_4, n is the moles of water per mole of CH_4 in the hydrate, R is the gas constant and T is the absolute temperature. a_w is the activity of water relative to pure liquid water (the a_w includes only the reduction of activity due to dissolved salts and dissolved methane; the pressure effect on the water activity is included in the last term). $f'_{1 \text{ atm}}$ is the dissociation fugacity of methane hydrate when the hydrostatic pressure is 1 atm and when a_w = 1 (that is, there is no dissolved methane). $f'_{1 \text{ atm}}$ cannot be measured for CH_4 hydrate because the dissociation pressure is 26 atm at 0°, and because the methane cannot be prevented from dissolving in the water. Nevertheless, this quantity is useful for various thermodynamic calculations. $f_{P \text{ atm}}$ is the dissociation fugacity at P_{hydro} (atm of hydrostatic pressure).

The meaning of this can be seen from the hypothetical experiment shown in Figure 10. The hydrostatic pressure is maintained by the piston. The cylinder contains pure water and CH_4 hydrate. The membrane allows CH_4, but not liquid water, to pass through it. Teflon is a semi-permeable membrane that has been used in a similar experimental arrangement [*Enns et al.* 1965]; Teflon allows water vapor through, so the pressure gage reading can be corrected for this to give the partial pressure of CH_4 in equilibrium with the hydrate. From the P_{CH_4}, the $f_{P \text{ atm}}$ can be calculated (and the reverse), using the appropriate value of f/P_{CH_4}.

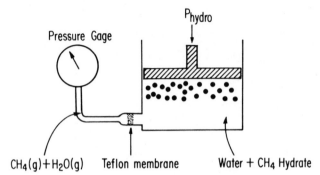

Fig. 10. Device for measuring hydrate dissociation pressures under hydrostatic pressures (P_{hydro}), which is maintained by the piston. The Teflon membrane allows CH_4 and water vapor to pass, but not liquid water. Methane hydrate being less dense than water is shown floating.

The value of $f'_{1\ atm}$ at 0° for CH_4 hydrate is calculated as follows: we take $n = 6.0$, $V_{H_2O} = 18.02$ cm^3 mole^{-1}, $V_h = 135.78$ cm^3 per mole of CH_4 in the hydrate (based on a 12.00 Å unit cell). The water activity term can be evaluated by

$$n \ln a_w = n \ln (1 - X_{CH_4}) = n \ln \left[1 - \frac{f_{CH_4}}{H} \right] \approx n \frac{f_{CH_4}}{H}$$

where H is Henry's law constant in atm per mole fraction (the pressure correction on Henry's law constant is omitted here because it is small, but is important in the solubility discussion below). The solubility of CH_4 at 0° is 2.55 x 10^{-3} molar per atm [*Claussen and Polglase, 1952; Wetlaufer et al. 1964*]. The corresponding value of H is 2.18 x 10^4 atm per mole fraction. For pure CH_4 hydrate, the hydrostatic pressure is the same as the dissociation pressure (26.04 atm). Using these numerical values, $f'_{1\ atm} = 23.59$ atm.

With this value of $f'_{1\ atm}$ we can calculate $f_{P\ atm}$ at any hydrostatic pressure and water activity from equation (1). As an example, we will calculate $f_{400\ atm}$ for water rather than for sea water. This corresponds to methane hydrate at the bottom of a fresh water lake of about 4000 meters depth. Putting the above constants into equation (1) and using $P_{hydro} = 400$ atm, we obtain $f_{400\ atm} = 39.71$ atm.

The pressure corresponding to this fugacity (f/P = 0.9012 for f = 39.71) is 44.06 atm. This is the pressure that will be measured by the pressure gage on the gas side of the Teflon membrane. As the hydrostatic pressure is 400 atm, a bubble of CH_4 cannot form even though the hydrate is stable.

Equation (1) shows that the hydrostatic pressure destabilizes the hydrate as the volume of the hydrate is greater than that of the water it forms on melting. The 400 atm of hydrostatic pressure increases the dissociation pressure by a factor of 1.69. This factor is small compared to the hydrostatic pressure increase, but large compared to the effect of the increased methane solubility (a factor of only 1.007) on lowering the water activity.

The amount of dissolved CH_4 can be calculated using the method of *Krichevsky and Kasarnovsky* [1935]. When Henry's law constant (H) is defined as f_{CH_4}/X_{CH_4}, instead of the usual expression P_{CH_4}/X_{CH_4}, we have

$$\ln(H_{P\,atm}/H_{1\,atm}) = \frac{\bar{V}_{CH_4}}{RT}(P_{hydro} - 1)$$

where $H_{P\,atm}$ and $H_{1\,atm}$ are f_{CH_4}/X_{CH_4} at P atm and 1 atm of hydrostatic pressure, respectively. $\bar{V}_{CH_4}$ is the partial molal volume of CH_4 dissolved in water. Taking $\bar{V}_{CH_4}$ as 35.0 cm^3 $mole^{-1}$ at 0° [*Glew, 1962b*], we have

$$H_{400\,atm}/H_{1\,atm} = 1.774$$

The amount of dissolved CH_4 at f_{CH_4} = 39.71 atm is 0.0571 molar compared to a calculated 0.101 molar if the pressure effect on Henry's law is not taken into account. It is to be noted that if water is saturated with methane at 1 atm fugacity (or pressure), increasing the hydrostatic pressure increases the fugacity of methane, but the fugacity increase is much smaller than the hydrostatic pressure increase, and bubbles of methane cannot form.

In the case where water is saturated with methane at 400 atm and bubbles of methane are present, the pressure of methane in the bubbles (i.e., the gas phase) is 400 atm. At this pressure f/P = 0.6210, which gives f = 245 atm, and the dissolved CH_4 = 0.352 molar. Methane hydrate is stable under these conditions and will form until the dissolved methane drops to the equilibrium value of 0.0571 molar. Thus, 0.295 moles per liter of methane will be converted to hydrate and will remove 1.770 moles per liter of liquid water to form the hydrate, or 3.2% of the liquid water.

The above calculation has been done for water. For sea water, the $\ln a_w$ term in equation (1) becomes important, and we can assume to a fair approximation that the various volumes and n remain unchanged. Taking the activity of water in sea water (at 1 atm pressure) as 0.9840, equation (1) predicts that $f'_{1\ atm}$ is raised 11.1% by the dissolved salts. The dissociation pressure is raised by approximately the same amount. At high hydrostatic pressures, the sea water figures are raised by about the same percentage over the pure water figures. This is in agreement with the data of *Deaton and Frost* [1946] and *Barduhn et al.* [1962]. The solubility of methane in sea water is about 20% less than pure water because of the salting-out effects [*Weiss, 1970*].

The effect of gases other than methane can be very important in stabilizing hydrates, particularly if the other gases form hydrates at low pressures. To a good approximation [*Miller, 1961a*], a mixed hydrate is stable (in equilibrium with water or ice and the gas phase) when

$$\frac{P_1}{P_1^{\circ}} + \frac{P_2}{P_2^{\circ}} = 1$$

where P_1 and P_2 are the partial pressures of gases 1 and 2 in equilibrium with the mixed hydrate, and P_1° and P_2° are the dissociation pressures of the pure hydrates of gases 1 and 2. At 0°, we have $P_1^{\circ} = 26.04$ atm for CH_4 and $P_2^{\circ} = 0.92$ atm for H_2S. Thus, if the gas phase contains 99% CH_4 and 1% H_2S, a hydrate will form when the total pressure is 20.5 atm ($P_{CH_4} = 20.3$ atm, $P_{H_2S} = 0.2$ atm). This is a reduction of 21% in the dissociation pressure caused by the presence of 1% H_2S.

REFERENCES

Barduhn, A. J., H. E. Towlson, and Y. C. Hu, The properties of some new gas hydrates and their use in demineralizing sea water, *AIChE, J., 8*, 176, 1962.

Barduhn, A. J., Desalination by crystallization processes, *Chem. Eng. Prog., 63*, 98, 1967.

Barrer, R. M., and W. I. Stuart, Nonstoichiometric clathrate compounds of water, *Proc. Roy. Soc., London, A243*, 172, 1957.

Bhatnagar, V. M., *Clathrate Compounds*, S. Chand and Co., New Delhi, 1968.

Chan, J. P., and W. F. Giauque, The entropy of $NH_3 \cdot 2H_2O$. Heat Capacity from 15 to $300^{\circ}K$, *J. Phys. Chem., 68*, 3053, 1964.

Claussen, W. F., Suggested structures of water on inert gas hydrates, *J. Chem. Phys.*, *19*, 159, 1951a.

Claussen, W. F., Erratum: suggested structures of water on inert gas hydrates, *J. Chem. Phys.*, *19*, 662, 1951b.

Claussen, W. F., and M. F. Polglase, Solubilities and structures in aqueous aliphatic hydrocarbon solutions, *J. Amer. Chem. Soc.*, *74*, 4817, 1952.

Davidson, D. W., Clathrate hydrates, in *Water: A Comprehensive Treatise*, edited by F. Franks, Plenum Press, New York (in press, vol. 5), 1973.

Davy H., On some of the combinations of oxy-muriatic gas and oxygen, and on the chemical relations of the principals to inflammable bodies, *Phil. Trans. Roy. Soc., London*, *101*, 1, 1811.

Deaton, W. M., and E. M. Frost, Jr., Gas hydrates and their relation to the operation of natural-gas pipe lines, *U.S. Bureau of Mines Mono., No. 8*, 1946.

Delsemme, A. H., and D. C. Miller, Physico-chemical phenomena in comets--II. Gas adsorption in snows of the nucleus, *Planet. Space Sci.*, *18*, 717, 1970.

Delsemme, A. H., and P. Swings, Hydrates de gas dans les noyaux cométaires et les grains interstellaires, *Ann. Astrophys.*, *15*, 1, 1952.

Delsemme, A. H., and A. Wegner, Physico-chemical phenomena in comets--I. Experimental study of snows in a cometary environment, *Planet. Space Sci.*, *18*, 709, 1970.

Din, F. (ed.), *Thermodynamic Functions of Gases*, vol. 3, pp. 64-67, Butterworths, London, 1961.

Donath, W. E., Purification of salt water, U.S. Patent 2,904,511, 1959, *Chem. Abst.*, *54*, 1779, 1960.

Enns, T., P. F. Scholander, and E. D. Bradstreet, Effect of hydrostatic pressure on gases dissolved in water, *J. Phys. Chem.*, *69*, 389, 1965.

Frank. H. S., The structure of ordinary water, *Science*, *169*, 635, 1970.

Frank, H. S., and M. W. Evans, Free volume and entropy in condensed systems III. Entropy in binary liquid mixtures; partial molal entropy in dilute solutions; structure and thermodynamics in aqueous electrolytes, *J. Chem., Phys.*, *13*, 507, 1945.

Frost, E. M., Jr., and W. M. Deaton, Gas hydrate composition and equilibrium data, *Oil Gas J.*, *45* (July 27), 170, 1946.

Glew, D. N., The gas hydrate of bromochlorodifluoromethane, *Can. J. Chem.*, *38*, 208, 1960.

Glew, D. N., Dehydration of solutions, U.S. Patent 3,058,832, 1962, *Chem. Abst.*, *57*, 16340, 1962a.

Glew, D. N., Aqueous solubility and the gas-hydrates. The methane-water system, *J. Phys. Chem.*, *66*, 605, 1962b.

Glew, D. N., Concentration of aqueous solutions through hydrate formation, French Patent 1,470,548., *Chem. Abst.*, *67*, 74956, 1967.

Goldberg, P., Free radicals and reactive molecules in clathrate cavities, *Science*, *142*, 378, 1963.

Gow, A. J., H. T. Ueda, and D. E. Garfield, Antarctic ice sheet: Preliminary results of first core hole to bedrock, *Science*, *161*, 1011, 1968.

Hafemann, D. R., and S. L. Miller, The clathrate hydrates of cyclopropane, *J. Phys. Chem.*, *73*, 1392, 1969a.

Hafemann, D. R., and S. L. Miller, The deuteriohydrates of cyclopropane, *J. Phys. Chem.*, *73*, 1398, 1969b.

Hagen, M., *Clathrate Inclusion Compounds*, Reinhold, New York, 1962.

Hammerschmidt, E. G., Formation of gas hydrates in natural gas transmission lines, *Ind. Eng. Chem.*, *26*, 851, 1934.

Hess, M., and G. E. Jones, Jr., Process and apparatus for separating water from an aqueous system, U.S. Patent 3,119,772, 1964, *Chem. Abst.*, *60*, 11750, 1964.

Hildebrand, D. L., and W. F. Giauque, Ammonium oxide and ammonium hydroxide. Heat capacities and thermodynamic properties from 15 to 300°K, *J. Amer. Chem. Soc.*, *75*, 2811, 1953.

Jeffrey, G. A., and R. K. McMullan, The clathrate hydrates, *Prog. Inorg. Chem.*, *8*, 43, 1967.

Jhaveri, J., and D. E. Robinson, Hydrates in the methane-nitrogen system, *Can. J. Chem. Eng.*, *43*, 75, 1965.

Kamb, B., A clathrate crystalline form of silica, *Science*, *148*, 232, 1965a.

Kamb, B., Structure of ice VI, *Science*, *150*, 205, 1965b.

Kamb, B., and B. L. Davis, Ice VII, the densest form of ice, *Proc. Nat. Acad. Sci. U.S.*, *52*, 1433, 1964.

Kauzmann, W., Some factors in the interpretation of protein denaturation, *Advan. Protein Chem.*, *14*, 1, 1959.

Kavanau, J. L., *Water and Solute Interactions*, Holden-Day, San Francisco, 1964.

Kingery, W. D. (ed.), *Ice and Snow*, MIT Press, Cambridge, Mass., 1963.

Kliorc, A., G. Fjeldbo, and B. L. Seidel, Mariners 6 and 7: Radio occultation measurements of the atmosphere of Mars, *Science*, *166*, 1393, 1969.

Klotz, I. M., Protein hydration and behavior, *Science*, *128*, 815, 1958.

Klotz, I. M., Noncovalent bonds in protein structure, *Brookhaven Symp. Biol.*, *13*, 25, 1960.

Krichevsky, I. R., and J. S. Kasarnovsky, Thermodynamical calculations of solubilities of nitrogen and hydrogen in water at high pressures, *J. Amer. Chem. Soc.*, *57*, 2168, 1935.

Lewis, J. S., The clouds of Jupiter and the NH_3-H_2O and NH_3-H_2S system, *Icarus*, *10*, 365, 1969.

Mak, T. C., and R. K. McMullan, Polyhedral clathrate hydrates X. Structure of the double hydrate of tetrahydrofuran and hydrogen sulfide, *J. Chem. Phys.*, *42*, 2732, 1965.

Makogon, Yu. F., F. A. Trebin, A. A. Trofimuk, V. P. Tsarev, and N. V. Cherskiy, Detection of a pool of natural gas in a solid (hydrated gas) state, *Dokl. Akad. Nauk SSSR (Earth Sci.)*, English Transl., *196*, 203, 1971.

Mandelcorn, L., Clathrates, *Chem. Rev.*, *59*, 827, 1959.

Marshall, D. R., S. Saito, and R. Kobayashi, Hydrates at high pressures: Part I. Methane-water, argon-water, and nitrogen-water systems, *AIChE J.*, *10*, 202, 1964.

McKoy, V., and O. Sinanoglu, Theory of dissociation pressures of some gas hydrates, *J. Chem. Phys.*, *38*, 2946, 1963.

Miller, S. L., The occurrence of gas hydrates in the solar system, *Proc. Nat. Acad. Sci. U.S.*, *47*, 1798, 1961a.

Miller, S. L., A theory of gaseous anesthetics, *Proc. Nat. Acad. Sci. U.S.*, *47*, 1515, 1961b.

Miller, S. L., Clathrate hydrates of air in Antarctic ice, *Science*, *165*, 489, 1969.

Miller, S. L., and W. D. Smythe, Carbon dioxide clathrate in the Martian ice cap, *Science*, *170*, 531, 1970.

Neugebauer, G., G. Münch, S. C. Chase, Jr., H. Hatzenbeler, E. Miner, and D. Schofield, Mariner 1969: Preliminary results of the infrared radiometer experiment, *Science*, *166*, 98, 1969.

Neugebauer, G., G. Münch, H. Kieffer, S. C. Chase, Jr., and E. Miner, Mariner 1969 infrared radiometer results: Temperatures and thermal properties of the Martian surface, *Astron. J.*, *76*, 719, 1971.

New York Times (Financial Section), U.S.-Soviet Gas Deal?, by Gene Smith, Jan. 2, p. 1, 1972.

Nye, J. F., The flow law of ice from measurements in glacier tunnels, laboratory experiments and the Jungfraufirn borehole experiment, *Proc. Roy. Soc., London, A219*, 477, 1953.

Olovsson, I., and D. H. Templeton, The crystal structure of ammonia monohydrate, *Acta Cryst., 12*, 827, 1959.

Ottawa Citizen, The, Large solid gas deposits in the polar regions, by I. Ilyinskaya, Oct. 16, p. 18, 1971.

Otto, F. D., and D. B. Robinson, A study of hydrates in the methane-propylene-water system, *AIChE J., 6*, 602, 1960.

Palin, D. E., and H. M. Powell, Hydrogen bond linking of quinol molecules, *Nature, 156*, 334, 1945.

Palin, D. E., and H. M. Powell, The structure of molecular compounds: Part V. The clathrate compounds of quinol and methanol, *J. Chem. Soc.* (118), 571, 1948a.

Palin, D. C., and H. M. Powell, The structure of molecular compounds: Part VI. The β-type clathrate compounds of quinol, *J. Chem. Soc.* (163), 815, 1948b.

Parker, A., Potable water from sea water, *Nature, 149*, 184, 1942.

Pauling, L., and R. E. Marsh, The structure of chlorine hydrate, *Proc. Nat. Acad. Sci. U.S., 38*, 112, 1952.

Powell, H. M., Clathrates, in *Non-Stoichiometric Compounds*, edited by L. Mandelcorn, p. 438, Academic Press, New York, 1964.

Siemons, W. J., and D. H. Templeton, The crystal structure of ammonium oxide, *Acta Crystallogr., 7*, 194, 1954.

Stackelberg, M. von, Feste gashydrate, *Naturwissenschaften, 36*, 327, 1949a.

Stackelberg, M. von, Feste gashydrate, *Naturwissenschaften, 36*, 359, 1949b.

Stackelberg, M. von, and H. Frübuss, Feste Gashydrate. IV. Doppelhydrate, *Z. Elektrochem., 58*, 99, 1954.

Stackelberg, M. von, and W. Jahns, Feste gashydrate. IV. Die Gitteraufweitungsarbeit, *Z. Elektrochem., 58*, 162, 1954.

Stackelberg, M. von, and W. Meinhold, Feste gashydrate III. Mischhydrate, *Z. Elektrochem., 58*, 40, 1954.

Stackelberg, M. von, and H. R. Müller, On the structure of gas hydrates, *J. Chem. Phys., 19*, 1319, 1951.

Stackelberg, M. von, and H. R. Müller, Feste gashydrate II. Struktur und Raumchemie, *Z. Elektrochem., 58*, 25, 1954.

Stoll, R. D., J. Ewing, and G. M. Bryan, Anomalous wave velocities in sediments containing gas hydrates, *J. Geophys. Res.*, *76*, 2090, 1971.

Takenouchi, S., and G. C. Kennedy, Dissociation pressures of the phase $CO_2 \cdot 5\text{-}3/4H_2O$, *J. Geol.*, *73*, 383, 1965.

van der Waals, J. H., and J. C. Platteeuw, Clathrate solutions, *Advan. Chem. Phys.*, *2*, 1, 1959.

Weiss, R. F., The solubility of nitrogen, oxygen and argon in water and seawater, *Deep-Sea Res.*, *17*, 721, 1970.

Wetlaufer, D. B., S. K. Malik, L. Stoller, and R. L. Coffin, Nonpolar group participation in the denaturation of proteins by urea and guanidinium salts. Model compound studies, *J. Amer. Chem. Soc.*, *86*, 508, 1964.

Whipple, F. L., A comet model. II. Physical relations for comets and meteors, *Astrophys. J.*, *111*, 464, 1951.

Williams, V. C., Hydrate-forming saline-water conversion project, U.S. Patent 2,974,102, 1961, *Chem. Abst.*, *55*, 17964, 1961.

REVIEW OF GAS HYDRATES WITH IMPLICATION FOR OCEAN SEDIMENTS

J.H. Hand, D.L. Katz and V.K. Verma

Department of Chemical Engineering
The University of Michigan
Ann Arbor, Michigan 48104

ABSTRACT

This paper presents a general review of gas hydrate knowledge, the utilization of such knowledge in the natural gas industry, some contemplated uses of gas hydrates in sea water desalinization, and discusses the existence of natural gas hydrates under the permafrost of the Northern Hemisphere. The theory of how gases enter the water phase and cause premature crystallization of water into an ice-like hydrate structure is presented. With this background, hydrate formation in ocean sediments is considered.

INTRODUCTION

When water and certain gases are mixed at high pressure, an ice-like solid precipitates from the system at temperatures well above the ice formation point. These unusual compounds, called gas hydrates, have been known and studied for over 160 years. Recently, with the discovery by *Stoll et al.* [1971] of the presence of methane hydrate in ocean sediments, a new area of interest in the general subject of gas hydrates arose. This paper presents a brief review of gas hydrate knowledge for the benefit of oceanographers and marine geologists who may be encountering the subject for the first time. Specifically, we will present experimentally determined hydrate formation conditions for pure and mixed gases combined with pure water or brine. The theory of hydrate formation will then be discussed. Next, some novel applications of hydrates will be presented and finally, with this background, the implications of this information for hydrate formation in ocean sediments will be considered.

179

EXPERIMENTAL EVIDENCE

A general review of the development of gas hydrate knowledge up to 1945 is given in a bibliography compiled by *Katz and Rzasa* [*1946*]. The earliest work of Davy in 1810 and *Faraday* [*1823*] on chlorine hydrates was continued, mainly by the French physical chemists, with *Villard* [*1888*] a prime investigator. His paper showed the conditions under which ethylene and methane (the primary gaseous constituent of marine sediments) would form hydrates. Figure 1 is a plot of the temperatures and pressures at which methane forms hydrates with pure water. Data to some 4000 psia, due to Villard, have been verified and extended by more recent investigations [*Kobayashi and Katz, 1949; Marshall et al. 1964*]. In the 1930's, the discovery that gas transmission lines were being plugged by natural gas-water hydrate formation spurred activity in the field. Figure 2 shows the hydrate forming conditions for the paraffinic hydrocarbon family. A full review of the work on natural gas hydrates undertaken until 1957 is presented in *Katz et al.* [*1959*].

The phase relations of a hydrate system can best be studied by two dimensional pressure-temperature diagrams such as Figures 1 to 4, which are sections, at constant composition, of the complete three dimensional cube. According to the Gibbs phase rule

(Degrees of Freedom = Number of Components - Number

of Phases + 2)

a single gas-water system has (4-Number of Phases) degrees of freedom. The system has five possible equilibrium phases

1) Gas saturated with water vapor, V
2) Water saturated with gas, L_1
3) Liquified gas saturated with water, L_2
4) Ice, I
5) Hydrate, H

If three phases are present, only one of the temperature or pressure conditions is needed to completely specify the system. Sets of pressure-temperature conditions for hydrate formation are represented by a hydrate curve on the phase diagram as shown in Figures 1 through 5. Consider the case of ethane in Figure 2. The line AB represents the $V-L_1-H$ hydrate curve. As pressure and temperature are increased along AB, point B is eventually reached. At B, the gas phase V begins to condense and four phases $V-L_1-L_2-H$ can be present. According to the phase rule, such a system has no degree of freedom. The point B is therefore unique and is called the quadruple point.

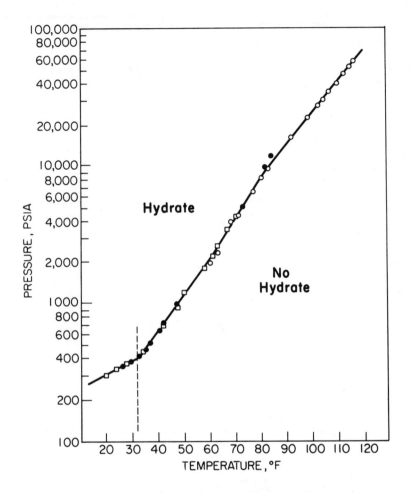

Fig. 1. Hydrate formation conditions for methane.

A similar quadruple point exists for propane, butane, and other gases whose critical point is above the ice point. Such quadruple points do not exist for methane or ethylene-water systems. At pressures greater than that of point B, gas phase V does not exist, giving rise to the L_1-L_2-H hydrate curve BC. Similarly at temperatures below the ice point, the absence of the L_1 phase gives rise to a V-H-I curve.

The L_1-L_2-H curves BC for ethane (and B'C' for propane) on Figure 2 are roughly vertical, indicating only a slight effect of pressure on the hydrate forming temperature above the quadruple point B (or B'). This lack of pressure dependence can be visualized

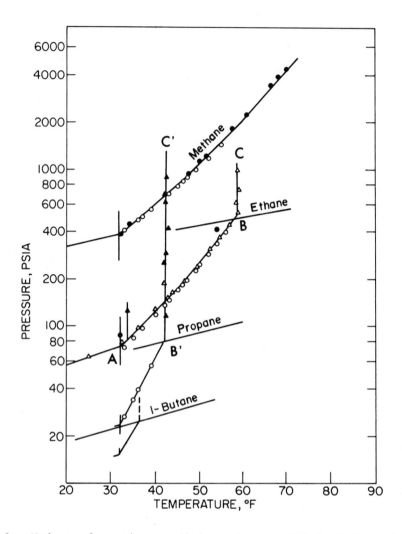

Fig. 2.   Hydrate formation conditions for paraffinic hydrocarbons.

through the thermodynamic Clapeyron's equation

$$\frac{dP}{dT} = \frac{1}{V}\frac{\Delta H}{\Delta V} \tag{1}$$

where ΔH = change in enthalpy during hydrate formation, and ΔV = change in volume during hydrate formation.

As ΔV from the liquid phase to the solid phase is zero or near zero, the L_1-L_2-H curve is almost vertical. The actual slope can be positive or negative, depending on the sign of ΔV.

The addition of another gas into the system increases the number of degrees of freedom, thus generating hydrate curves for specific fixed composition mixtures as shown on Figures 3 and 4. Each of these hydrate curves will end in a quadruple point if the gas phase can be condensed.

Introduction of another gas generally has dramatic effect on the hydrate forming conditions. *Deaton and Frost* [1946] showed this effect by adding ethane and propane in measured concentrations to methane. Propane has the most pronounced effect on the hydrate forming conditions for methane as shown in Figure 3. Only 1% propane in a methane-propane mixture can reduce the pressure at which methane hydrate forms by nearly 40%. Similar effects for the carbon dioxide-methane system, which should be of interest in the marine environment, are shown in Figure 4 [*Unruh and Katz, 1949*]. Using this knowledge, charts have been prepared showing the effect of gas gravity on hydrate formation [*Katz et al. 1959*]. Vapor-solid equilibrium ratios have been empirically established in order that one can predict, from the composition of the gas, the temperature-pressure relationships for hydrate formation when water is present [*Katz et al. 1959*].

Sodium chloride, alcohols or other 'anti-freezes' lower the temperature at which hydrates will form just as these substances lower the freezing point of water [*Katz et al. 1959*]. The hydrate curves at various salt concentrations shown in Figure 5 illustrate the effect of salt water as compared to fresh water on the hydrate forming conditions for methane. As it is of interest to know the condition at which methane might form hydrate with sea water, a curve has been drawn on this chart as an interpolation between pure water and the concentrations measured [*Kobayashi et al. 1951*].

THEORY OF HYDRATE FORMATION

Using X-ray diffraction techniques, *Stackelberg and Müller* [*1951*] established that hydrates generally crystallize in one of two different crystal structures, depending upon the size of hydrate forming solute molecules. Structure I consists of 46 water molecules, and has 2 cavities of 7.88 Å average diameter and 6 larger cavities of 8.6 Å average diameter. Structure II consists of 136 water molecules, and has 16 cavities of 7.82 Å average diameter and 8 larger cavities of 9.46 Å average diameter. Structure I is shown schematically in Figure 6.

The two lattice structures I and II are metastable for pure

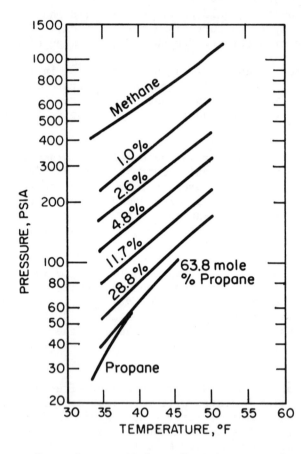

Fig. 3. Hydrate formation conditions for mixtures of methane and propane.

water and would not exist above 32°F. The hydrate forming solute molecules, which may originate from a gaseous phase or a liquid hydrocarbon phase, enter the water and are trapped in the cavities stabilizing the structure and allowing crystals to form at temperatures above the ice point. The interaction between the encaged solute molecules and the surrounding network of water is not chemical in nature, but is due to the van der Waals forces similar to that found between adjacent molecules in liquids. The structure of the hydrate formed depends only on the size of the 'guest' solute molecules. Hydrates of Structure II are formed only by solute molecules that are too large to fit in the cavities of Structure I. If only the large cavities of Structure II hydrate are occupied, the formula for these hydrates can be written as (solute)$\cdot$136/8 (H_2O) or (solute)$\cdot$17H_2O. Similarly, for Structure I, the formula for full

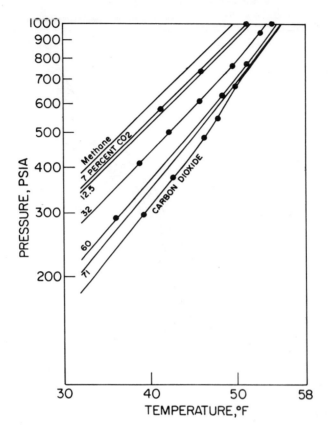

Fig. 4. Hydrate formation conditions for mixtures of methane and carbon dioxide.

occupation can be written as (solute)·46/8 (H_2O) or (solute)·5-3/4 (H_2O). Experimentally, we find these hydrate numbers of 5-3/4 and 17 for Structures I and II respectively are upper limits, because a partial occupation of available cavities can stabilize the crystal structure under suitable conditions. Figure 7 taken from *Katz et al.* [*1959*] shows the size of various hydroformer molecules and the type of hydrates they can form.

In addition to those qualitative remarks, the thermodynamic properties of hydrate crystals can be quantitatively represented by statistical thermodynamics due to their regular crystal structure. The basic equation describing hydrate formation in mixed gas systems was derived by *van der Waals and Platteeuw* [*1959*] and is shown below. Specifically, this equation shows that the chemical potential of water in the vacant, metastable lattice is lowered by

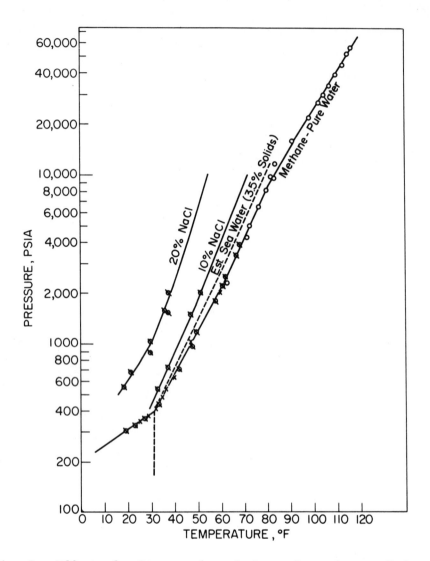

Fig. 5. Effect of salt on methane hydrate formation conditions.

inclusion of the hydrate former due to an entropic effect caused by 'mixing' the solute with the lattice.

$$\mu_W^H - \mu_W^\beta = kT \sum_i \nu_i \ln (1 - \sum_j y_{j_i}) \tag{2}$$

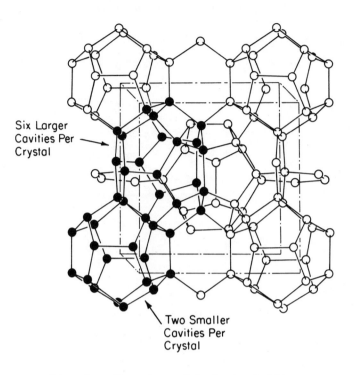

Six Larger Cavities Per Crystal

Two Smaller Cavities Per Crystal

Fig. 6. Structure I hydrate lattice.

where

μ_W^H = chemical potential of water in hydrate crystal

μ_W^β = chemical potential of water in empty lattice structure

k = Boltzmann constant

T = temperature

ν_i = number of cavities of type i per mole of water

y_{j_i} = fraction of cavities of type i occupied by molecules of type j

The fractional occupancy y_{j_i} can be calculated by evaluating the energy of interaction between the solute and the hydrate lattice.

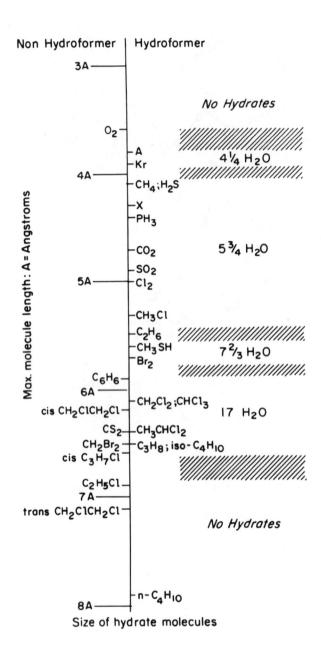

Fig. 7.　Guest size, hydrate structure relation.

$$y_{j_i} = \frac{C_{j_i} f_j}{(1 + \sum_\ell C_{\ell_i} f_\ell)} \tag{3}$$

where f_j = the fugacity of solute j in hydrate, and C_{j_i} = a type of Langmuir adsorption constant, which may be evaluated once a potential function for the water-solute interaction is defined. The traditional approach is to ignore the fact that discrete water molecules form the cage by effectively 'smearing' the water molecules into a spherical surface. Then a spherically symmetric cell potential $W(r)$ can be written. The Langmuir constant can then be calculated from

$$C_{j_i} = \frac{4}{kT} \int_0^\infty \exp\left[-\frac{W(r)_{j_i}}{kT}\right] r^2 dr \tag{4}$$

This constant has been evaluated using the Lennard-Jones 12-6 symmetric potential function, the Lennard-Jones 28-7 symmetric potential function [*McKoy and Sinanoglu, 1963*] and the Kihara potential [*Parrish and Prausnitz, 1972*]. Recently, the discrete nature of the water cage has been taken into account using a Monte-Carlo approach [*Tester et al. 1972*]. Results are similar to the approximate theory.

Equations (1), (2), and (3) can be used to predict hydrate forming conditions given the values of $(\mu_W^H - \mu_W^\beta)$ and C_{j_i}. *Parrish and Prausnitz* [*1972*] have presented a stepwise procedure to calculate the dissociation pressure for hydrate using $(\mu_W^H - \mu_W^\beta)$ as that of a reference hydrate and using C_{j_i}'s that have been back calculated to fit the best available data. Their procedure is claimed to work well for gas-water-hydrate systems for one solute or for multi-solute systems. The procedure may not, however, predict hydrate forming conditions for liquid hydrocarbon phase-aqueous phase-hydrate systems. Much more experimental data are necessary to correlate Langmuir constants for liquid-liquid-hydrate systems.

NOVEL USAGE AND OCCURRENCE OF HYDRATES

We conclude our review with a brief discussion of the use of hydrates in demineralizing sea water and the existence of hydrates in the earth and in ocean sediments.

Desalination

The fact that hydrates, like ice, crystallize with the exclusion of ions led to a flurry of work [*Knox, 1961; Braduhn et al. 1960, 1963*] in the early 1960's in the investigation of hydrate formation at temperatures above 32°F, as an alternate to freezing sea water to free it of mineral content. By using propane, isobutane or the Freons, one can reduce the refrigeration requirement in the form of a higher temperature level at which a solid will precipitate with a hydrate forming substance present. Processes involving hydrate formation are presently economically marginal, but additional work with other solutes may make the process more attractive.

Hydrates in the Earth

The announcement made in early 1970 that a hydrate field had been found in Siberia was followed by a study showing the depths at which one would expect hydrates in the earth [*Katz, 1971*]. Figure 8 shows the temperature gradient in the earth at Prudhoe Bay on the North Slope of Alaska. Note that there is 2000 feet of permafrost and the temperature at that depth is 32°F. A hydraulic column 2000 feet deep would have a pressure of some 860 psi at its base. Because methane or natural gases can form hydrates at temperatures much higher than 32° at this pressure, a substantial 'frozen gas' field could exist. Accordingly, one can cross-plot the hydrate formation conditions and the earth temperature-pressure conditions on a common graph and find the depth at which one should expect to find hydrates, depending on the composition of the gas and the reservoir.

Natural gas dissolved in a crude oil could well enter the water phase by solution and hence form hydrates when no gas phase is present [*Katz, 1972*]. Early work done with the methane-propane systems showed that one could measure the hydrocarbon-liquid-water-hydrate formation temperature and pressure in the form of a locus of quadruple points for the system. It would follow that the same measurements could have been made in the absence of the vapor phase. It would also follow that any mechanism for bringing those constituents that form hydrates into solution in water could be used to form 'gas hydrates'. No gas phase is required. It should be noted that carbon tetrachloride forms hydrates and, of course, is thought of as a liquid because its boiling point is considerably above room temperature.

FORMATION OF HYDRATES IN OCEAN SEDIMENTS

As mentioned above [*Stoll et al. 1971*], gases emanating from ocean sediments cores brought to the surface may be due to gas release from hydrate decomposition at atmospheric pressure. If someone

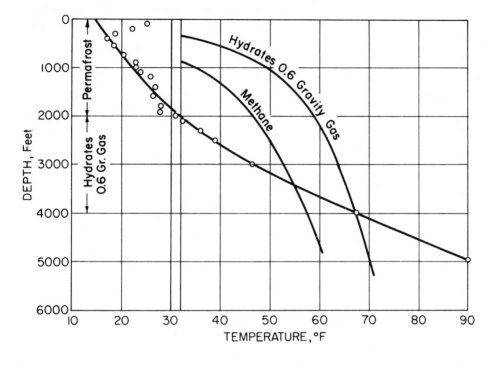

Fig. 8. Hydrate formations in Arctic reservoirs.

would have asked those working in the field of gas hydrates whether methane would form hydrate at the conditions in water corresponding to temperatures and pressures of the curve in Figure 1, the answer would have been affirmative.

In fact, methane, ethane, propane, isobutane will also form hydrates in deep sea conditions, as will carbon dioxide, carbon monoxide, and hydrogen sulfide. Hydrocarbons larger than isobutane do not appear to form hydrates, primarily because they are too large to fit in the biggest solute cages in the metastable lattice. Hydrogen and helium also do not form hydrates, either because they are too small to be trapped by the water cages or because their polarizability is so low that van der Waals interaction between hydrogen or helium and water is too small to significantly stabilize the crystal.

With regard to ocean sediments, it would be helpful to know how hydrocarbons are formed and enter the water phase. Reviews of such hydrocarbon formation by *Breger and Whitehead* [*1951*], *Breger* [*1960*], and *Erdman* [*1960*] were consulted for guidance. It would appear that the precise mechanism for formation is not understood. Indeed,

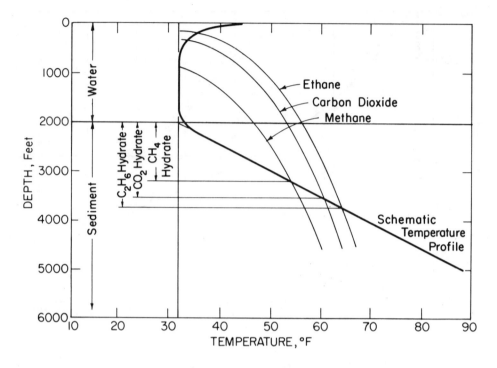

Fig. 9.   Hydrate formation in ocean sediment.

in view of the complexity of the system involved and the varied conditions in sediments, it may be that several mechanisms are operating.

Degradation products of proteins, carbohydrates, and lipids should certainly be investigated. On decomposition, proteins can be expected to liberate CO_2, H_2S, H_2O and NH_3. The first three products are known to be hydrate formers. Carbohydrates yield CO_2, alcohols, fats, and water. Here we recognize one hydrate former, CO_2, and one substance, alcohol, which would inhibit hydrate formation if present in sufficient concentration. Finally, lipids are decomposed into fatty acids, sterols, and alcohols, which all can be further attacked.

In advancing their radioactive mechanism, Berger and Whitehead experimentally investigated the interaction between fatty acids and alpha particles. The products of the reactions consisted of the hydrate formers CO_2, CO and lower hydrocarbons and the non-hydrate former H_2. *Zobell* [*1947*] suggests that much of the hydrogen may be consumed by a bacterially catalyzed reaction with CO_2 to form methane and water. *Breger* [*1960*] has also proposed that fatty acids may be converted to more reactive acids by bacterial action with subsequent decarboxylation. In addition, there is the possibility of direct bacterial production of hydrocarbons, especially methane [*Zobell, 1961*].

This small sampling of the literature reveals a vast array of proposed hydrocarbon formation reactions. Regardless of the actual mechanism, if hydrate forming molecules are produced or migrate into contact with water under suitable conditions, a hydrate can be formed. Figure 9 shows a cross-plot of ocean temperature, pressure, and hydrate forming conditions for a conventional natural gas-sea water system. Note that methane hydrate can be formed to a depth of 1100 feet in sediment. This correlates quite well with measurements of the depth of anomalous reflectors reported by the U.S. Navy. As final proof of the existence of oceanic hydrates, a pressure core barrel should be developed to take and maintain samples at *in situ* conditions for subsequent evaluation.

REFERENCES

Barduhn, A. J., H. E. Towlson, and Y. C. Hu, *United States Office of Saline Water Progress Rep. 44,* 1960.

Barduhn, A. J., H. E. Towlson, and Y. C. Hu, The properties of some new gas hydrates and their use in demineralizing sea water, *AIChE J., 8,* 176, 1963.

Breger, I. A., Diageneses of metabolites and a discussion of the origin of petroleum hydrocarbons, *Geochim. et Cosmochim. Acta, 19,* 1960.

Breger, I. A., and W. L. Whitehead, Radioactivity and the origin of petroleum, *Proc. 3rd World Petrol. Congr., Sec. I,* pp. 421-427, 1951.

Deaton, W. M., and E. M. Frost, Gas hydrates, *U.S. Bureau of Mines Mono. No. 8,* 1946.

Erdman, J. G., Petroleum--its origin in the earth, in *Fluids in Subsurface Environments,* edited by A. Young and J. E. Galley, *Amer. Ass. Petrol. Geol.,* 1960.

Faraday, M., On the condensation of several gases into liquids, *Phil. Trans. 22,* 189, 1823.

Katz, D. L., Depths to which frozen gas fields may be expected, *J. Petrol. Tech., April,* 419, 1971.

Katz, D. L., Depths to which frozen gas fields may be expected - footnotes, *J. Petrol. Tech., May,* 557, 1972.

Katz, D. L., and M. J. Rzasa, *Bibliography for Physical Behavior of Hydrocarbons under Pressure and Related Phenomena,* J. W. Edwards, Inc., 1946. (Presently available from University Microfilms, Ann Arbor, Mich.)

Katz, D. L., D. Cornell, R. Kobayashi, F. H. Poettmann, J. A. Vary, D. R. Elenbaas, and C. F. Weinaug, *Handbook of Natural Gas Engineering,* McGraw-Hill, New York, 1959.

Knox, W. G., *et al.,* The hydrate process, *Chem. Eng. Prog., 57,* 66, 1961.

Kobayashi, R., and D. L. Katz, Methane hydrate at high pressure, *Trans. AIME, 186,* 66, 1949.

Kobayashi, R., *et al.,* Gas hydrate formation with brine and ethanol solutions., *Proc. 30th Ann. Conv. Nat. Gasoline Ass. of Amer.,* pp. 27-31, 1951.

Marshall, D. R., S. Saito, and R. Kobayashi, Hydrates at high pressures, Part I. methane-water, argon-water and nitrogen-water, *AIChE J., 10,* 202, 1964.

McKoy, V., and O. Sinanoglu, Theory of dissociation pressure of some gas hydrates, *Chem. Phys., 38,* 2946, 1963.

Parrish, W. R., and J. M. Prausnitz, Dissociation pressures of gas hydrates formed by gas mixtures, *I.E.C. Proc. Des. Develop., 11,* 26, 1972.

Stackelberg, M. von, and H. R. Müeller, On the structure of gas hydrates, *J. Chem. Phys., 19,* 1319, 1951.

Stoll, R. D., J. Ewing, and G. M. Bryan, Anomalous wave velocities in sediments containing gas hydrates, *J. Geophys. Res., 76,* 2090, 1971.

Tester, J. W., R. L. Bivins, and C. C. Herrick, Use of Monte Carlo in calculating the thermodynamic properties of water clathrates, *AIChE J., 18,* 1220, 1972.

Unruh, C. H., and D. L. Katz, Gas hydrates of carbon dioxide-methane mixtures, *Trans. AIME, 186,* 83, 1949.

van der Waals, J. H., and J. C. Platteeuw, Clathrate solutions, *Adv. in Chem. Physics,* vol. 2, Interscience, New York, 1959.

Villard, M., Dissolution des liquides et des solids dans les gaz, *Compt. Rend., 106,* 453, 1888.

ZoBell, C. E., Microbial transformation of molecular hydrogen in marine sediments with particular reference to petroleum, *Amer. Ass. Petrol. Geol. Bull, 31,* 1709, 1947.

ZoBell, C. E., Contributions of bacteria to the origin of oil, *Proc. 3rd World Petrol. Congr., Sec. I,* pp. 414-420, 1951.

OCCURRENCE OF NATURAL GAS HYDRATES IN SEDIMENTARY BASINS

Brian Hitchon

Alberta Research
Edmonton, Alberta
Canada

ABSTRACT

Clathrates are a special variety of inclusion compound in which the guest molecules fit into separate spherical or nearly spherical chambers within the host molecule, and when the host molecule is water and the guest molecules are largely gases or liquids with low boiling points found in natural gas, the clathrates are termed natural gas hydrates. They are solid compounds, resembling ice or wet snow in appearance, and form both below and above the freezing point of water under specific PT conditions. The water molecules form pentagonal dodecahedra, which can be arranged into two different structures, leaving interstitial space in the form of either tetrakaidecahedra or hexakaidecahedra. Methane and hydrogen sulfide can be accommodated in all the spaces, ethane and carbon dioxide can fit in both the tetrakaidecahedra and the hexakaidecahedra, but propane and isobutane fit only in the hexakaidecahedra. Normal butane, pentane, and hexane are not known to form hydrates. PT diagrams describing the initial conditions for hydrate formation indicate that, relative to methane, all common components of natural gas (except nitrogen and the rare gases) raise the hydrate formation temperature, propane and ethane being the most effective. The presence of dissolved salts in the water, or nitrogen and rare gases in the natural gas, depresses the temperature of initial hydrate formation.

The most likely way to produce natural gas hydrates in sedimentary basins is through a reduction of temperature, rather than an approach to lithostatic pressures, and the most pertinent situation is that found in regions with relatively thick permafrost sections. Sedimentary basins with extensive areas of relatively thick, continuous permafrost, which may contain potentially commercial occurrences of natural gas hydrates, are limited to the Arctic slope petroleum province of Alaska, the Mackenzie Delta and Arctic Archipelago of Canada, and the northern portion of the West Siberian basin and the Vilyuy basin of the U.S.S.R. Natural gas hydrates have only been well documented in the Messoyakha field in the West Siberian basin, which contains 14 trillion ft^3 of potentially recoverable natural gas, partly in hydrate form. That portion of the field with natural gas hydrates contains 54% higher reserves than would be expected based on the assumption that the reservoir rocks were filled with free gas. Conclusive evidence of natural gas hydrates in wells drilled in the arctic regions of North America is lacking, although it is anticipated they will be found. There is need for careful monitoring of wells drilled in permafrost regions, because the nature of natural gas hydrates makes them difficult to detect unless proper precautions are taken. The technology for drilling in natural gas hydrate zones and for the optimum recovery of gas reserves in the form of natural gas hydrates is in its infancy, and more research is required.

INTRODUCTION

Inclusion compounds are physical combinations in which one component fits into a cavity in the other. There are no ordinary chemical bonds between the atoms of the host molecule and those of the guest molecule. Binding between the molecules is by means of dispersion forces. There are six different forms of inclusion compound, depending on the molecular architecture of the host structures and the shapes of the cavities they enclose. The six forms are nicely illustrated by *Brown* [*1962*] and are: 1. Clathrates, produced when guest molecules fit into separate spherical or nearly spherical chambers within the host molecule; 2. canal complexes, formed when the host is a crystal lattice having tubular cavities; 3. layer complexes, in which there are alternating layers of guest and host; 4. molecular sieves, in which there are interconnected chambers and passageways; 5. intramolecular hollow space complexes, formed when the host is a large molecule containing a concavity or depression in which the guest molecule is accommodated; and 6. linear polymer complexes, produced when guest molecules fit into the tube formed by a pipelike host molecule. Only one variety of clathrate is of concern in this study.

Clathrates may be considered as compounds in which guest molecules are enclosed within a cagelike structure of the host molecule and from which they cannot escape until the structure is destroyed. A variety of substances may be used as host molecules, including quinol and water. The substances with which water forms clathrates are largely gases, or liquids with low boiling points, and these clathrate complexes are generally termed gas hydrates.

Gas hydrates are solid compounds, resembling ice or wet snow in appearance, which form between water and a variety of substances, both below and above the freezing point of water, and under specific conditions of temperature and pressure. They behave like solutions of gases in a solid metastable water lattice. They were discovered by Humphry Davy in 1811, and the early classical contributions to our knowledge of gas hydrates by Faraday, Roozeboom, Tammann, Villard, and de Forcrand have been reviewed by *Schröder* [*1927*]. However, it was not until quite recently that their molecular architecture was confirmed by use of molecular models and X-ray diffraction techniques [*Stackelberg, 1949; Claussen, 1951a; Stackelberg and Müller, 1951; Claussen, 1951b; Pauling and Marsh, 1952; Jeffrey and McMullan, 1963*]. Although bromine gas hydrate crystallizes in the tetragonal class [*Jeffrey and McMullan, 1963; Allen and Jeffrey, 1963*], most gases that form gas hydrates, and certainly all the gases that occur in natural gas and that also form gas hydrates, produce hydrates crystallizing in the cubic system in one of two distinct structures.

In ordinary ice the water molecules are linked together, through hydrogen bonds, into rings with six water molecules. Ice crystallizes in the hexagonal system, and in ordinary ice (ice I) the hydrogen-bonded framework does not contain any chambers large enough for occupancy by molecules other than those of helium or hydrogen. By contrast, in hydrates, the water molecules are linked together into rings with five molecules, and these pentagonal planes further join to form dodecahedra (Figure 1). The length of the hydrogen bond is 2.75 Å in hydrates, essentially the same as in ordinary ice (2.76 Å). Unlike ordinary ice, the dodecahedra of hydrates (mean free diameter $\simeq$ 5.1 Å) are sufficiently large to accommodate small molecules like argon (3.08 Å), krypton (3.38 Å), xenon (3.80 Å), methane (4.06 Å), and hydrogen sulfide (4.20 Å). The values in parenthesis are the largest van der Waals diameters for the respective components. But even more significant is the fact that space cannot be filled completely by any packing arrangement of dodecahedra, and this leaves some interstitial space between the dodecahedra. The dodecahedra pack like atoms in diamond type or body-centered type cubic crystals, each of which gives rise to a slightly different arrangement of the space between the dodecahedra. In the body-centered type of packing (Structure I), the interstitial space is defined by 24 water molecules at the corners of a tetrakaidecahedron (mean free diameter $\simeq$ 5.8 Å), a figure with 2 hexagonal faces and 12 pentagonal faces (Figure 1). In the diamond type of packing (Structure II) the interstitial space is defined by 28 water molecules at the corners of a hexakaidecahedron (mean free diameter $\simeq$ 6.7 Å), a figure with 4 hexagonal faces and 12 pentagonal faces (Figure 1). Molecules like argon, krypton, xenon, methane, and hydrogen sulfide can be accommodated in all the spaces, carbon dioxide (5.10 Å) and ethane (5.43 Å) can fit in both the tetrakaidecahedra and the hexakaidecahedra, but propane (6.63 Å) and iso-butane fit only in the hexakaidecahedra. Recent laboratory studies have shown that normal

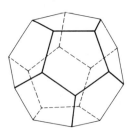

DODECAHEDRON

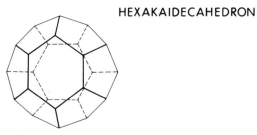

HEXAKAIDECAHEDRON

TETRAKAIDECAHEDRON

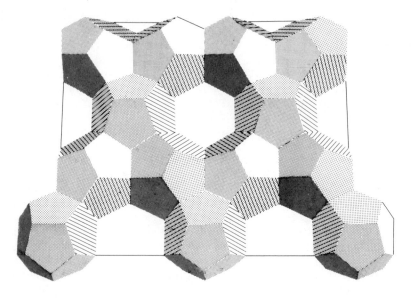

Fig. 1. Models of a dodecahedron, a hexakaidecahedron and a tetrakaidecahedron. Below, is a portion of the Structure I gas hydrate
model showing the arrangement of one layer of dodecahedra, and the
resulting interstitial space in the shape of tetrakaidecahedra
[*Brown, 1962*]. Only one-half of the tetrakaidecahedra are shown.

TABLE 1. Properties of Gas Hydrate Lattices

Property	Structure I	Structure II
Unit cell size (Å)	12	17
Number of dodecahedra	2	16
Number of tetrakaidecahedra	6	-
Number of hexakaidecahedra	-	8
Number of water molecules per unit cell	46	136
Per cent cell size increase over ice I	16	18
Stoichiometric formulas:		
All spaces filled	X. $5\text{-}3/4H_2O$	X. $5\text{-}2/3H_2O$*
Only 'kaidecahedra filled	X. $7\text{-}2/3H_2O$	X. $17H_2O$

*Never observed, and probably unstable.

butane does not form hydrates (U.S. Office of Saline Water, Research and Development Progress Report No. 292), and neither do pentane and hexane. Table 1 lists some of the properties of these two packing types of gas hydrate lattices. The type of packing depends to some extent on the size of the guest molecules present, and although double hydrates and mixed hydrates are known, discussion of them is not relevant to this study. These and other aspects of gas hydrates have been reviewed by *Katz et al.* [*1959*], *Bhatnager* [*1962*], *Byk and Fomina* [*1968*], *Jeffrey* [*1969, 1972*], and *Davidson* [*1971*].

It is the intention of this paper to evaluate the occurrence of natural gas hydrates in sedimentary basins, first, through a review of pertinent pressure-temperature-composition data for components of natural gas and associated formation fluids, and second, by an examination of the limits for the occurrence of natural gas hydrates set by this review, with particular reference to the distribution of permafrost in the northern hemisphere. Finally, sedimentary basins producing hydrocarbons, or those potentially productive, within

Arctic permafrost regions will be noted, known natural gas hydrate occurrences will be described, and the known and potential commercial natural gas hydrate accumulations evaluated.

PRESSURE-TEMPERATURE-COMPOSITION LIMITATIONS

The most convenient way to view the limits of hydrate formation for the components in natural gas is by means of pressure-temperature diagrams showing the initial conditions for the formation of hydrates for the pure components (or mixtures) in binary systems comprising the component (or mixture) and water. Shown in Figure 2 is that portion of the phase diagram that is pertinent to this discussion, the stippled region indicating where hydrates exist. The line ECF represents the equilibrium between liquid and vapor for the component in question, and lies at a slightly higher pressure than for the pure component due to the presence of water. In the region bounded by DCF, the condensed or liquid phases of both the component and water are present, whereas in the region GBCF, vapor (containing a small amount of water) and water (containing some dissolved component) are present. With excess water present, cooling of either the liquid component and water, or the vapor component and water, results in the formation of a two-phase hydrate and water system. The line BH deviates from 0°C to the extent that the dissolved component and pressure lower the freezing point, and cooling below 0°C causes the liquid water to change to ice. Within the region ABG, the two phases present are ice, and vapor of the component. Points B and C are termed quadruple points, with four phases present at each point. The line CD represents equilibrium between the hydrate, water, and liquid component and is approximately vertical. Of concern to this study is that portion of the phase diagram, both below and above the freezing point of water, in which the hydrate phase is present.

The region of the phase diagram in which hydrates occur for each of the pure components with water is illustrated in Figure 3. In the left-hand diagram, pertinent information on hydrate formation by hydrocarbons has been superimposed onto one phase diagram, and in the right-hand diagram information is given on hydrate formation by the non-hydrocarbon components of natural gas. Please refer to *Katz et al.* [1959] and *Jhaveri and Robinson* [1965] for more specific details and control points used to construct these diagrams. From the relative positions of the lines for the initial conditions for the formation of pure hydrates, it is clear that, relative to methane, all components except nitrogen (and the rare gases) will raise the hydrate formation temperature, whereas nitrogen (and the rare gases) will depress the hydrate formation temperature. However, not all the components are equally effective in changing the hydrate formation temperature, as may be seen by an examination of Figure 4. The five phase diagrams in Figure 4 illustrate the effect of adding ethane, propane, nitrogen, carbon dioxide, and hydrogen sulfide,

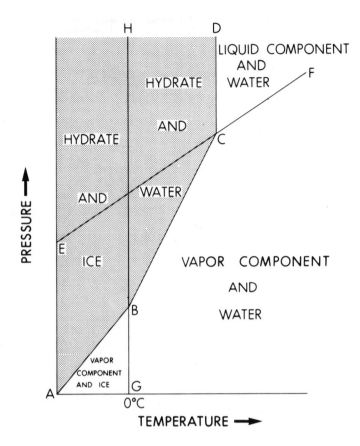

Fig. 2. Phase diagram illustrating the fields of hydrate formation (stippled). Modified after *Wilcox et al.* [*1941*].

separately, to methane, and the resulting changes in the initial conditions for the formation of hydrates at 1, 5, 10, and 50 molal % of the added component. For a given added molal %, propane has the greatest effect on changing the initial conditions for the formation of hydrates, then hydrogen sulfide and ethane, with carbon dioxide and nitrogen having the least, but in the case of nitrogen, opposite effects. Propane and ethane will therefore be most effective

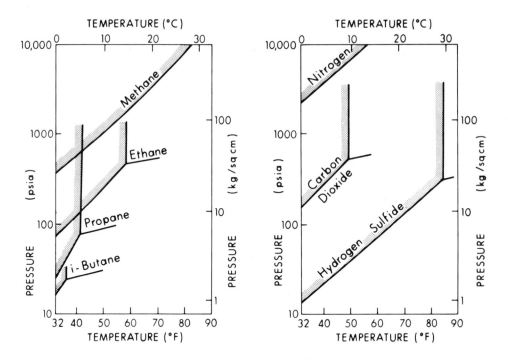

Fig. 3. Phase diagrams showing the initial conditions of hydrate formation for individual hydrocarbons and water (left-hand diagram) and for non-hydrocarbon gases and water (right-hand diagram).

in changing the initial conditions for the formation of hydrates in natural gases in view of the paucity of hydrogen sulfide in many natural gases. The phase diagrams shown in Figure 4 represent the simplest systems. More complex ones with three or more components have been studied by *Katz* [1945] and by *Robinson and Hutton* [1967]. One study that is particularly pertinent is that of *Kobayashi et al.* [1951], which examined the effect of dissolved salts in the water on the initial conditions for the formation of hydrates. The effect of adding various amounts of sodium chloride to the aqueous phase with which methane is forming the hydrate is shown in Figure 5. Salt added to the water lowers the temperature at which the hydrates melt or form, following the colorful description by *Katz* [1971], "as in making ice cream in the country freezer".

In order to evaluate the possibility of the occurrence of natural gas hydrates in sedimentary basins, it is necessary to relate the geothermal, geopressure, and salinity gradients of a typical sedimentary basin to the phase diagrams presented in Figures 3, 4, and 5. This has been done in Figure 6 in which the temperature, hydrostatic pressure (fresh water), and salinity data for the Alber-

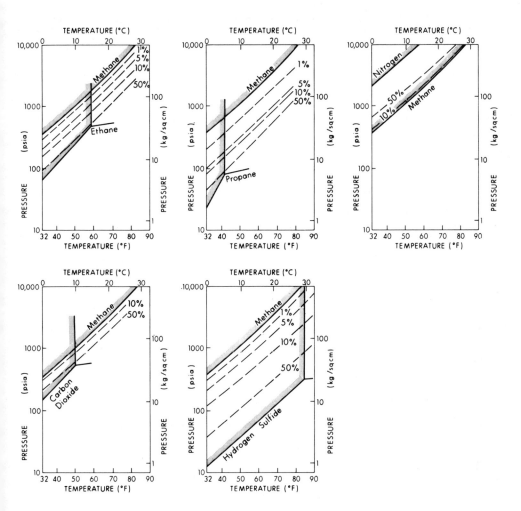

Fig. 4. Phase diagrams illustrating the relative effects of the separate addition of ethane, propane, nitrogen, carbon dioxide and hydrogen sulfide to methane-water systems with respect to the initial conditions of hydrate formation. Percentage data refer to molal amount of the added component relative to methane. Diagrams modified after *Katz et al.* [*1959*], *Jhaveri and Robinson* [*1965*], *Unruh and Katz* [*1949*], and *Noaker and Katz* [*1954*].

ta portion of the western Canada sedimentary basin are plotted against depth, in the conventional manner, on the left-hand diagram, and on the right-hand diagram the same temperature and pressure gradients (with indicated depths) are transposed to a pressure temperature phase diagram showing the initial conditions for hydrate

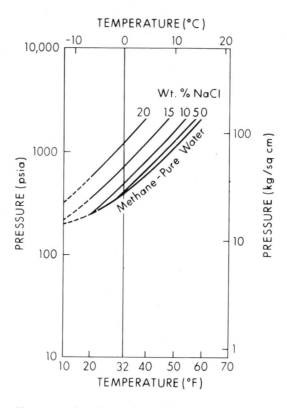

Fig. 5. Phase diagram showing the effect of adding sodium chloride to the water on the initial conditions of hydrate formation for methane. Modified after *Kobayashi et al.* [*1951*].

formation for methane, ethane, and propane. Studies of fluid flow in the western Canada sedimentary basin [*Hitchon, 1969a,b*] have shown that conditions are not hydrostatic, and that pressure gradients are greater than hydrostatic for fresh water and are associated with updip flow out of the deeper portions of the basin. This fact serves to raise the PT-depth curve in the right-hand diagram, although the effect will be minimal at shallower depths. Reductions of temperature near to, or below, the freezing point of water would bring the PT-depth curve well within the region of hydrate formation for the shallower depths, but would have little effect at greater depths. This observation is not new, having been presented previously by *Katz* [*1971*], although the method of study here may be different. It does indicate, however, that a reduction of temperature, rather than an approach to lithostatic pressures, is more likely to lead to conditions conducive to hydrate formation. The most obvious environment where such situations may prevail are

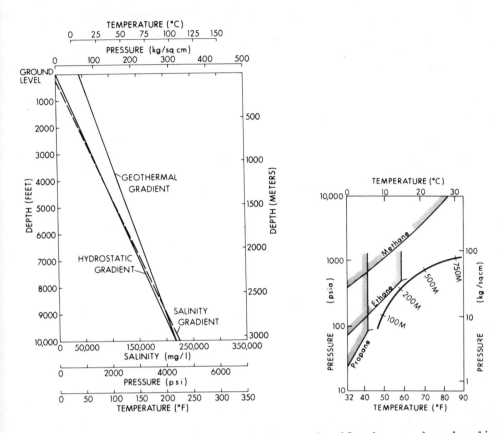

Fig. 6. Relation of geothermal, hydrostatic (fresh water) and salinity gradients for the Alberta portion of the western Canada sedimentary basin to depth (left-hand diagram) and to the initial conditions for hydrate formation (right-hand diagram).

found in permafrost regions, and the distribution of these regions will now be reviewed.

DISTRIBUTION OF PERMAFROST

Baranov [1964] has identified three so-called cryogenic regions of the Earth. These are: 1) areas of short-lived and irregular seasonal freezing of the soil; 2) areas of regular seasonal freezing of the soil, and 3) regions of permafrost and ice cover. We are concerned only with category 3 in this study. Permafrost is defined solely on a temperature basis, and is applied to regions of the Earth that are continuously at 0°C, or below, on a season to season basis. Permafrost is present in the two circumpolar regions, as well as in temperate and tropical zones near and above the

TABLE 2. World Distribution of Permafrost*

Region	Area Occupied (km^2 x 10^6)	
Northern Hemisphere:		
U.S.S.R.	11	
Canada	5.7	
Greenland	1.6(?)	
Alaska	1.5	
Mongolian Peoples Republic	0.8	
China (without Tibet)	0.4	
Total Northern Hemisphere		21.0
Southern Hemisphere:		
Antarctica	13.5(?)	13.5
Total for Both Hemispheres		34.5

*After *Baranov* [*1964*].

permanent snow line. About 23% of the total land area of both hemi-
spheres (149 x 10^6 km^2) is occupied by permafrost; Table 2 shows
the breakdown by regions [*Baranov, 1964*]. These data may change
significantly if it is shown that permafrost does not exist under
the Greenland and Antarctic ice sheets, or is of limited extent.
Permafrost areas in Tibet, the Himalayas, Hindu Kush, Iceland, northern Scandinavia, the Alps, and the individual permafrost islands in
the Cordillera of the North and South American continents were excluded because they were difficult to delineate. Nevertheless, this
total area of permafrost must be considered as the possible maximum
area. No account was taken by Baranov of permafrost underlying the
oceans, although he indicates its extent in the Kara, Laptev, and
East Siberian Seas [*Baranov, 1964*, Figure 24].

Although scientists and engineers working in all permafrost regions should be cognizant of the possible presence of natural gas
hydrates, only those permafrost regions of Canada, Alaska and the
U.S.S.R. are of more immediate commercial concern. Even within these
regions, continuous and discontinuous zones of permafrost may be
distinguished. In the continuous zone, permafrost lies beneath all
land areas, but is absent directly beneath large water bodies, which
provide a sufficient heat reservoir to keep the subjacent bottom
sediments unfrozen [*Mackay, 1972a*]. The zone of discontinuous

permafrost comprises a complex mosaic of frozen and unfrozen ground; indeed, it is so bewilderingly complex that we will confine our attention to the region of continuous permafrost, while recognizing that the boundary between the zones of continuous and discontinuous permafrost is ill-defined. Conclusions drawn with respect to the zone of continuous permafrost do not apply to all areas in the zone of discontinuous permafrost, where permafrost is present, because the ground temperatures in the latter zone are very close to 0°C, whereas the ground temperatures in the continuous zone are considerably lower. During the Pleistocene glaciation the region of continuous permafrost was much more extensive than today, and hence the possibility for the occurrence of natural gas hydrates was much greater. In this context, *Makogon et al.* [*1972*] have measured the abrupt decrease in solubility of methane in water at PT conditions corresponding to the hydrate equilibrium, and have noted the possibility that some of the natural gas fields in the West Siberian basin and Yakutia were formed as a result of decomposition of hydrates during the recession of the permafrost and decrease in thickness of the zone of hydrate formation. It is interesting to speculate how we might recognize gas deposits that now exist as free gas but which were in the hydrate form during the Pleistocene glaciation.

The distribution of continuous permafrost in the northern hemisphere is shown in Figure 7. For Canada, the information was obtained from *Brown* [*1967a,b; 1970*], for Alaska, from *Ferrians* [*1965*], and for the U.S.S.R. from *Baranov* [*1964*], *Brown* [*1967b*], *Tricart* [*1970*], and *Mackay* [*1972a*]. Regions of offshore permafrost have been delineated by *Baranov* [*1964*] and *Mackay* [*1972a,b*]. Within the zone of continuous permafrost, large areas are underlain by non-sedimentary rocks, mainly Precambrian metamorphic rocks. Only the three rectangular map-areas in Figure 7 contain sedimentary basins which either produce hydrocarbons or are potentially productive, and each region will now be described in turn, the known natural gas hydrate occurrences noted, and their commercial potential evaluated.

ARCTIC SLOPE PETROLEUM PROVINCE, ALASKA

The southern limit of continuous permafrost in Alaska is approximately coincident with the southern edge of the Brooks Range (Figure 8). Only isolated measurements of the thickness of the permafrost are available [*Ferrians, 1965*], and these have been indicated in Figure 8. Permafrost thicknesses range from about 70-100 meters near the southern limit of continuous permafrost to generally more than 300 meters on the Arctic coastal plain, and a record 600 meters at the Prudhoe Bay oilfield [*Howitt, 1971*]. Physiographically, the region includes the Brooks Range, the Arctic foothills, and the Arctic coastal plain. Petroleum potential of the Brooks Range is not very encouraging because of the lithology and complex structure, although possible reservoir beds are present at

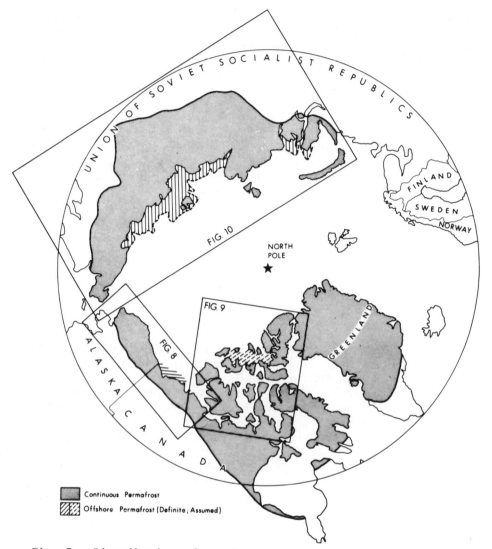

Fig. 7. Distribution of continuous permafrost in the northern hemisphere [after *Baranov, 1964; Brown, 1967a,b, 1970; Ferrians, 1965; Mackay, 1972a,b;* and *Tricart, 1970*].

the surface and do contain traces of petroleum residues [*Miller et al. 1959*]. No test wells have been drilled in this region.

The Arctic foothills, Arctic coastal plain, and the contiguous offshore basins to the edge of the continental shelf at about 200 meters of water depth comprise the Arctic slope petroleum province. During the decade from 1945 to 1955 the U.S. Navy Department

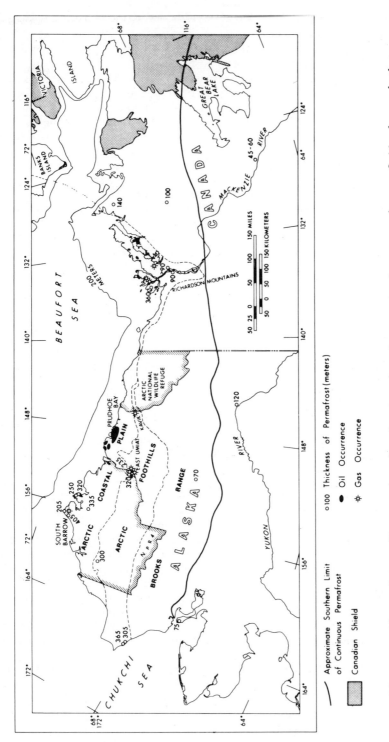

Fig. 8. Oil and gas discoveries in the Arctic slope petroleum province of Alaska and the Mackenzie Delta, Canada.

drilled 37 test wells and 45 core holes on 18 structures located within Naval Petroleum Reserve No. 4. Three oil fields and two gas fields were discovered with total reserves estimated at about 130 million barrels of oil and 370-900 billion ft^3 of natural gas. All are shut-in except the South Barrow gas field, which is utilized for local consumption. In the early 1960's, geological and geophysical exploration was carried out by the oil industry in the region between Naval Petroleum Reserve No. 4 and the Arctic National Wildlife Refuge. A few exploratory wells were drilled, which resulted in the discovery, in 1964, of the East Umiat gas field by BP Exploration East Umiat No. 1. This initial phase of exploration finally confirmed the petroleum potential of the Arctic slope petroleum province when Atlantic Richfield Prudhoe Bay No. 1 discovered the 10 billion barrels of oil and 26 trillion ft^3 of natural gas in the Prudhoe Bay field in 1968. Since then, three smaller oilfields have been discovered to the west of Prudhoe Bay, and the Kavik gas field was found in the Arctic foothills by Pan Am Kavik No. 1 in 1969.

There have been no published reports of the occurrence of natural gas hydrates in the Arctic slope petroleum province of Alaska. Unpublished and unconfirmed information indicates that cores taken in the permafrost zone and recovered to the surface, still under conditions of reservoir temperature and pressure, contain much greater quantities of natural gas than would normally be expected under reservoir PT conditions, a fact strongly indicative of the presence of natural gas hydrates *in situ* in the permafrost. Even beneath the permafrost zone conditions are still conducive for the occurrence of natural gas hydrates, and *Katz* [*1971*] has shown that at Prudhoe Bay, with 600 meters of permafrost, methane may be expected as a gas hydrate between 600 and 1035 meters, and a natural gas of 0.6 gravity would be in hydrate form between 600 and 1220 meters. The formation of hydrate would reduce the gas pressure if water does not move in to maintain reservoir pressure. One might therefore speculate that, in part, the low pressure in the 1140-1245 meter-deep reservoir discovered in Upper Cretaceous sandstones in Arco-West Sak River St. 1, in which the reservoir pressure was so low that the 18°-API gravity oil failed to flow to the surface [*Adams, 1972*], was the result of depletion of the gas cap through the formation of hydrates; it should be noted that gas hydrates can form from gases dissolved in liquids and do not require a free gas phase, and thus hydrates in crude oil or condensate reservoirs are possible [*Katz, 1972*]. Both these Alaskan observations concerning the possible occurrence of natural gas hydrates require confirmation. However, in view of the demonstrated presence of natural gas hydrates in the U.S.S.R. under similar conditions, there seems little reason to doubt the presence of natural gas hydrates under suitable PT conditions in the shallower oil and gas reservoirs in the Arctic slope petroleum province of Alaska.

MACKENZIE DELTA, CANADA

The Mackenzie Delta is part of the Arctic coastal plain and the entire delta lies within the zone of continuous permafrost (Figure 8). The few measurements of permafrost thickness that are available for that part of Canada are shown in Figure 8 and were obtained from *Brown* [*1967a*] and Judge (personal communication). The apparent thinness of most of the permafrost in the Mackenzie Delta, relative to that of the contiguous Arctic coastal plain of Alaska, is probably due to the effect of the Mackenzie River, its distributaries, and the many large lakes present in the delta region, all of which act as heat reservoirs and keep the bottom sediments unfrozen. *Johnston and Brown* [*1964*] have described the deep thawed zone beneath a typical lake in the Mackenzie Delta.

There has been considerable speculation regarding the existence of permafrost in Alaskan and Canadian offshore regions, including theoretical studies by *Lachenbruch* [*1957*], and extrapolations to offshore regions based on measurements made along the coast [*Brewer, 1958; Lachenbruch et al. 1966*]. The existence of permafrost beneath the sea floor of the southern Beaufort Sea was first established in 1970 by CSS Baffin and CSS Parizeau, from the discovery of submarine pingoes in both the eastern and western portions of the Beaufort Sea by means of side-scan sonar [*Anonymous, 1971*], and by the recovery of icy sediments and ice from some bore holes of the Ocean Floor Sampling Program of the Arctic Petroleum Operators Association of Calgary, Alberta. This latter program has been described in detail by *Mackay* [*1972b*], who concludes that permafrost is widespread in the southern Beaufort Sea, except where sea water temperatures are modified by large rivers such as the Mackenzie. However, in the Mackenzie Delta offshore region, where the mean annual sea bottom temperatures near Tuktoyaktuk are either close to or above zero, the permafrost is most certainly remnant in nature, and results from the rapid transgression of the sea over an area that previously had up to 300 meters of permafrost (Judge, personal communication). Thicknesses of permafrost up to 360 meters still exist in some nearshore areas. Inasmuch as sea bottom water temperatures in the deeper waters of the Queen Elizabeth Islands are generally below $-0.5°C$, then strictly by definition, permafrost should probably underlie a substantial portion of the Canadian arctic waters, although in fact the high salinity and lack of a history involving fresh water probably precludes the presence of ice in the bottom sediments. In the shallower waters between the islands of Queen Elizabeth Islands, where salinities are lower and the ice-sheets deposited much glacial till, permafrost thicknesses of up to 75 meters might be found under certain conditions in waters as deep as 250 meters. Close to shorelines, with land surface temperatures of about $-16°C$, essentially brackish or fresh water, and with possible freezing of the ice to the bottom mud in the winter, several hundreds of meters of permafrost might exist close to the shore (Judge, personal communication).

It may be assumed that similar conditions prevail in the Alaskan offshore regions of the Beaufort Sea.

No exploratory wells for oil and gas were drilled in the Mackenzie Delta region prior to 1965. Since that time there have been three oil discoveries and three gas discoveries, all made in the period 1970-1972. There have been no reports of the occurrence of natural gas hydrates, nor would they seem likely to occur in shallow regions adjacent to large bodies of water. However, the suggestion of *Mackay* [*1972b*] that permafrost thicknesses up to 400 meters may exist in some nearshore areas would imply that under these special circumstances natural gas hydrates could be expected in shallow reservoirs in these regions.

ARCTIC ARCHIPELAGO, CANADA

The Canadian Arctic Archipelago (Figure 9) comprises three physiographic regions. The Arctic coastal plain forms the western border of the Arctic Archipelago with the Arctic Ocean and extends from Banks Island to north of Ellef Ringnes Island, and is part of the same physiographic region as the Mackenzie Delta and the Arctic coastal plain of Alaska. The islands south of Viscount Melville Sound and Lancaster Sound comprise the Arctic lowlands, while those to the north form the upland and mountainous Innuitian region. All the Arctic Archipelago lies within the zone of continuous permafrost and, as *Mackay* [*1972b*] and Judge (personal communication) have suggested, permafrost probably underlies a substantial portion of the shallower waters between the islands. Judge (personal communication) has supplied the measurements of permafrost thickness shown in Figure 9. *Brown* [*1972*] has calculated from a mean annual air temperature of about -23°C that permafrost thicknesses in excess of 1000 meters may exist in the interior of Baffin and Ellesmere Islands, particularly at high topographic elevations.

Although commercial hydrocarbon accumulations may be present in the cover of sedimentary rocks overlying the Canadian Shield in the Arctic lowlands, there has been little drilling activity in the area, and most of the present exploration effort is directed towards the Sverdrup basin in the Innuitian region. The Sverdrup basin comprises the central lowland core of the Innuitian region, and is bounded on the south by the Parry Island and Cornwallis fold belts, on the east by the Central Ellesmere fold belt, and on the north and northwest by the Northern Ellesmere fold belt and the Arctic coastal plain, respectively.

The first wildcat well drilled for hydrocarbons in the Arctic Archipelago was Dome Petroleum Limited, Winter Harbour No. 1, which was spudded in 1961 near Winter Harbour on the south side of Melville Island, and subsequently abandoned. Two more wells were drilled and abandoned in the period up to December 1967, when Panarctic Oil Ltd.

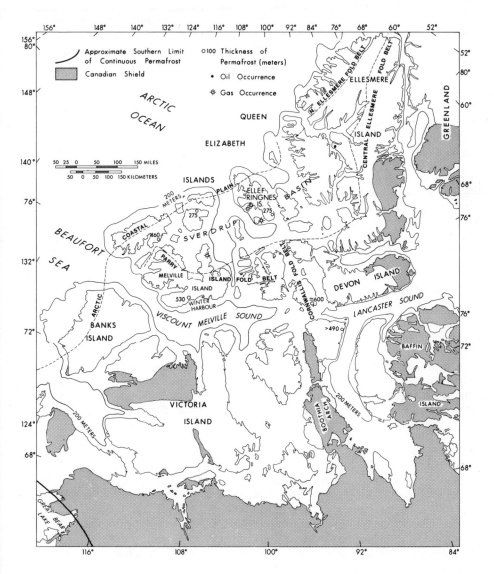

Fig. 9. Oil and gas discoveries in the Arctic Archipelago, Canada.

announced its extensive exploration program. Since that time, almost sixty wells have been drilled in the Canadian Arctic Archipelago to the end of 1972, resulting in four major discoveries of natural gas and one significant discovery of crude oil.

Little information has been released concerning the various discoveries made in the Arctic Archipelago, and there have been no reports of the occurrence of natural gas hydrates. One interesting point is the report (Allan Bryant, Dome Petroleum Limited, personal

communication) of a small gas flow of 61,000 ft^3/day from the Upper
Devonian Hecla Formation in the interval 319-727 meters in Dome
Petroleum Limited, Winter Harbour No. 1, in relation to the known
permafrost thickness at Winter Harbour of 530 meters (Judge, personal communication). The well was originally drilled to 3823 meters
and subsequently plugged back to 727 meters. The gas flow was obtained during swabbing operations in open hole between the end of
surface casing at 319 meters and the plug at 727 meters. The presence of natural gas hydrates is strongly indicated on the basis of
the available information. In some parts of the Arctic Archipelago,
gas has been encountered in the permafrost section at depths as
shallow as 3 meters, and in one instance an air drill discovered
shallow gas at about 15 meters, and the resulting fire destroyed a
seismic rig [*APOA, 1972*]. This shallow gas is probably the so-called
'marsh gas', generated by the anaerobic decay of organic material,
and may be similar to that described by *Mackay* [*1965*] from gas-domed
mounds in permafrost on Kendall Island in the Mackenzie Delta. It
is probably essentially only methane, and examination of Figure 6
indicates that unless an abnormally high pressure had built up (equivalent perhaps to 250 meters hydrostatic depth), these occurrences
are not likely to be in the form of gas hydrates. However, they do
point out the need for careful consideration of the potential occurrences of natural gas hydrates when drilling in permafrost regions
where thick permafrost is anticipated. Sudden release of pressure
during drilling could cause hydrate breakdown and the subsequent
release at high pressures of unexpectedly large quantities of gas,
causing casing failure, or expansion of drill cuttings and cores due
to expulsion of gas originally held in hydrate form.

U.S.S.R.

The approximate southern limit of continuous permafrost in the
U.S.S.R. is shown in Figure 10, based on maps by *Baranov* [*1964*],
Brown [*1967b*], *Tricart* [*1970*], and *Mackay* [*1972a*]. Obviously, considerable difficulty has been experienced in attempts to delimit
the southern boundary of continuous permafrost using U.S.S.R. data,
and *Brown* [*1967b*] has accepted a permafrost thickness of 60-100
meters at this boundary in Canada, and 250-300 meters at this boundary in the U.S.S.R. Although this may not be entirely satisfactory
from the point of view of those studying permafrost, the boundaries
are generally acceptable for this study, and we will consider only
a few U.S.S.R. hydrocarbon occurrences falling outside the zone of
continuous permafrost, to make this study equally comprehensive for
both North America and the U.S.S.R. *Baranov* [*1964*] has compiled a
map showing thickness of the permafrost in the U.S.S.R. in both the
continuous and discontinuous zones. Within the continuous permafrost
zone, thicknesses vary from 250-300 meters at the southern boundary
(by definition, *Brown* [*1967b*]) to over 500 meters in the northern
portions of the Yamal and Taymyr peninsulas, on the mainland

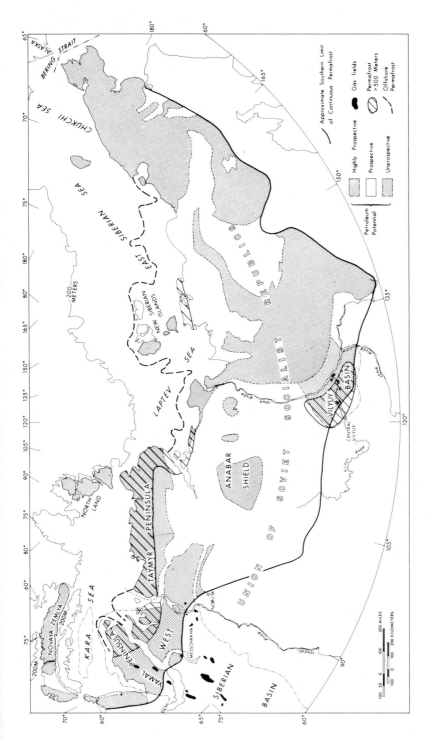
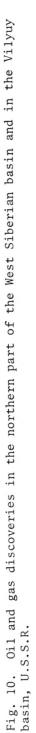

Fig. 10. Oil and gas discoveries in the northern part of the West Siberian basin and in the Vilyuy basin, U.S.S.R.

immediately south of the New Siberian Islands, and in central Yakutia near the confluence of the Lena and Vilyuy rivers (Figure 10).
It is not clear from Baranov's map whether the New Siberian Islands,
North Land and the northern half of Novaya Zemlya fall within the
region in which permafrost exceeds 500 meters, but in any case they
are generally rated unprospective with respect to hydrocarbons and
will not be considered further. The regions of offshore permafrost
have been delineated from *Baranov* [*1964*], but they may be overesti-
mated, according to U.S.S.R. sources quoted by *Mackay* [*1972a*].

   In 1967 the Ministry of Geology of the U.S.S.R. published a
map on the scale 1:5,000,000 titled "Map of oil and gas bearing
prospects U.S.S.R.", which classified the entire country, and some
offshore areas, according to three broad groups: highly prospective, prospective, and unprospective. There were many subdivisions
of these three groups that need not concern us, and thus only these
three major groups are shown in Figure 10 for those regions lying
within the zone of continuous permafrost. The map also includes a
highly prospective region in the Kara Sea. Within the map-area of
Figure 10 there are only two regions classified as highly prospective, and both contain important hydrocarbon reserves, almost solely
gas fields. They are the northern portion of the West Siberian
basin and the Vilyuy basin in central Yakutia, which are separated
by the unprospective Anabar shield. The large oil reserves in the
southern portion of the West Siberian basin lie just outside the
map-area. Although there has been some exploratory drilling within
the zone of continuous permafrost, but outside the two regions classified as highly prospective, the results have been generally un-
satisfactory with only minor discoveries of oil and gas. The location of major gas fields in the West Siberian and Vilyuy basins has
been taken from the above cited map and from *Eremenko et al.* [*1972*],
but regrettably, the author must make the same statement as *Eardley*
[*1972*], namely, that the number and location of oil and gas fields
in the U.S.S.R. is only approximately correct because available
maps and reports are not consistent. Nevertheless, the magnitude
of the gas reserves in the northern part of the West Siberian basin
and the Vilyuy basin may be judged from the fact that of the 79
world fields with the largest gas reserves, 20 are found in the West
Siberian basin and 2 in the Vilyuy basin [*Halbouty et al. 1970*].
Four out of the top five are from the West Siberian basin and nine
out of the top twenty are from this and the Vilyuy basins. The ultimate recoverable gas reserves of the top four fields in the West
Siberian basin are almost equal to the combined reserves of the next
15 fields, and three of these fields are also from the West Siberian
basin.

   In December 1969 the U.S.S.R. Committee for Inventions and Discoveries announced that significant reserves of natural gas hydrates
had been discovered in the U.S.S.R. The first report in the western
literature was that of *Chersky and Makogon* [*1970*], and although no

specific gas fields were mentioned, the regions cited in which the reserves were located were the northern portion of the West Siberian basin and the Vilyuy basin. *Makogon et al.* [*1971*] have detected about 30 pools that could contain natural gas hydrates, but only provide full details for the Messoyakha field, 275 km west of Noril'sk, and name the upper productive zones of the Central Vilyuy gas field in Yakutia, and the Dzhangostskaya and Ulakhan Yuryakh pools as among the 30 containing natural gas hydrates. The author has been unable to determine the location of the two latter pools, but the Messoyakha and Central Vilyuy fields are shown in Figure 10. The Central Vilyuy field is the 19th in the world in order of ultimate recoverable reserves (15.9 trillion ft^3) and the Messoyakha is the 20th, with ultimate recoverable reserves of 14.0 trillion ft^3 [*Halbouty et al. 1970*]. *Makogon et al.* [*1972*] provide minor additional information on the Yakutia occurrences.

Makogon and Medovskiy [*1969*] have compiled five maps showing the depth from the surface of the 0°, 5°, 10°, 15°, and 20° isotherms for essentially the entire region within the zone of continuous permafrost. They found that the most important factors controlling the depth of the isotherms were climatic and geographical conditions, the hydrodynamic flow pattern within the strata and the relief of the crystalline basement. Reduced mineralization of formation waters (presumably mainly in recharge areas, though they did not state this) was often associated with a decrease in temperature. Based on these five maps, they have drawn up a master map showing the maximum depth of hydrate formation, assuming uniform hydrostatic pressure and similar equilibrium parameters of hydrate formation throughout the map-area. Depths range from about 500 meters near the southern end of the Yamal Peninsula, to over 1200 meters in the Vilyuy basin, and over 1600 meters towards the lower reaches of the Lena River.

The following description of the natural gas hydrate occurrence at the Messoyakha gas field is taken entirely from *Makogon et al.* [*1971*]. The Messoyakha anticlinal structure is 19 km long and 12.5 km wide, and the Albian-Cenomanian reservoir has a vertical closure of 84 meters. The reservoir is a 76-meter thick sandstone sequence with 16-38% porosity (average 25%), 29-50% residual water content (average 40%), and a permeability ranging from several millidarcies to several hundred millidarcies. The reservoir lies at about 800-900 meters depth, and at that location the permafrost is 450 meters thick. Total dissolved solids of the formation water are 5000 mg/l and the natural gas contains 98.5% methane, 0.5% heavier hydrocarbons, and 1.0% nitrogen and inert gases. Reservoir pressure is 78 kg/cm^2 and the reservoir temperature ranges from 8-12°C. Pressure and temperature profiles based on actual measurements are shown in Figure 11, together with the temperature profile for initial hydrate formation from laboratory study of cores. At the reservoir pressure of 78 kg/cm^2, the temperature for initial hydrate formation is 10°C, and hence that portion of the reservoir lying at

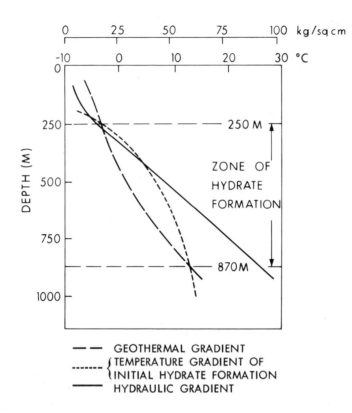

Fig. 11. Pressure and temperature relations in the Messoyakha field, West Siberian basin, U.S.S.R. [after *Makogon et al. 1971*].

depths shallower than the 10°C isotherm contains natural gas hydrates and that portion lying deeper than the 10°C isotherm contains free natural gas. The intersection of the geothermal and initial hydrate formation gradients at 250 meters and 870 meters (Figure 11) indicates the limits within which natural gas hydrates can exist. As might be expected, natural gas production was much higher from wells where the production perforations were in the free gas zone. Injection of methanol into test wells perforated in the hydrate zone increased their productivity by one order of magnitude. Calculations showed that because the natural gases were in the hydrate form, the reserves were 54% higher in that portion of the field with natural gas hydrates than would be expected on the assumption that the reservoir rocks were filled by free gas.

This complete description of the Messoyakha natural gas hydrate

occurrence confirms the presence of these compounds in nature.
With permafrost thicknesses exceeding 1065 meters in northern Yakutia [*Chersky and Makogon, 1970*] and low geothermal gradients, many
of the 30 possible occurrences noted by *Makogon et al.* [*1971*] will
doubtlessly be confirmed to contain natural gas hydrates.

DETECTION OF NATURAL GAS HYDRATES *IN SITU*

It is particularly difficult to distinguish gas hydrates by
purely chemical means, and conclusive proof of the *in situ* presence
of natural gas hydrates in a reservoir will only be possible by
X-ray diffraction analysis of a core containing the hydrates that
has been recovered to the surface at reservoir temperature and pressure.   Indirect evidence includes that cited for the Arctic Slope
petroleum province of Alaska, and that given by *Makogon et al.*
[*1971*] for the Messoyakha gas field of the West Siberian basin where
the amount of natural gas held in the reservoir is considerably
greater than that expected on the assumption that the reservoir rocks
were filled by free gas, and the PT conditions in the reservoir fall
within the PT field for hydrate formation.   Enhanced production of
natural gas through breakdown of the *in situ* hydrates by increasing
the bottom hole temperature, by artificial reduction of the bottom
hole pressure, or by injection of selected solvents, might also be
regarded as indirect evidence of the presence of natural gas hydrates.

Gas hydrates are probably present within the sediments in many
parts of the oceans, as first suggested by *Sokolov* [*1966*] (according
to *Vasiliev et al.* [*1973*]), and further studied by *Stoll et al.*
[*1971*], *Makogon* [*1972*], and by *Trofimuk et al.* [*1973*], and may be
detected, tentatively, by their specific seismic reflections [*Stoll,
1974*].   It is doubtful if on-shore seismic techniques can be refined
to do more than distinguish the base of the permafrost, and thus the
possible presence of natural gas hydrates in strata within and beneath the permafrost would remain undetected.   However, in regions
in which information on the depth to the base of the permafrost is
known (either through seismic studies or drilling), there remains
the possibility of direct determination of natural gas hydrates by
well logging techniques. *Chersky et al.* [*1972*] have suggested that
both mud logging and core gasometry (gas detection) can be used to
detect stratigraphic units containing natural gas hydrates, and they
provide caliper, resistivity, and SP logs for the productive interval at the Messoyakha gas field.   In addition, they also indicate
problems that can occur when drilling, perforating, and stimulating
wells in natural gas hydrate zones.   In part, these problems reflect
the fact that heat is required to decompose gas hydrates and that
decomposition takes place most rapidly in conditions of gas undersaturation and large thermal or pressure gradients [*Trebin et al.
1966*], all of which occur when the stable subsurface conditions are
disturbed by drilling.   Little information is available on the

physical properties of natural gas hydrates. Precision measurements have been made of the density of propane hydrate (d = 0.88 g cm^{-3}), and this seems to be the only density measurement available for the natural gas hydrates of interest to this study. G.A. Jeffrey (personal communication) has pointed out that the theoretical maximum density of the stoichiometric gas hydrate can be calculated from the X-ray crystallographic data for specific components of natural gas, as can the theoretical density of the water framework without the guest molecules (approximately 0.79 g cm^{-3}). However, because the hydrates in nature are non-stoichiometric and partial undersaturation with natural gas may exist, judgment must be exercised with respect to the composition of the natural gas and the degree of undersaturation. Corrections should also be made for pressure, and as an approximation the compressibility of ice could be used. It is the author's opinion that with the expected contrast in density between natural gas hydrates and formation water, with its variable content of dissolved salts, there is the possibility of refining and modifying the present techniques used for the Formation Density Log to distinguish between free gas and natural gas hydrate. Another possibility is to use the Borehole Compensated Sonic-Gamma Ray log, which *Howitt [1971]* has shown to be excellent for determining the thickness of permafrost, and could be modified and used alone, or in combination with other logs, to detect natural gas hydrate deposits.

PRODUCTION OF NATURAL GAS HYDRATE DEPOSITS

The only published information regarding the actual production of natural gas from natural gas hydrate deposits is that of *Makogon et al. [1971]* and *Chersky et al. [1972]*, in which they report the injection of methanol into test wells that were perforated in the hydrate zone in the Messoyakha field of the West Siberian basin, with resultant one order of magnitude increase in productivity. Experience in the U.S.S.R. with the formation of hydrate plugs at the bottom of the well bore, especially during the initial stages of production, suggests the use of a 30% calcium chloride solution, with the addition of 10% methanol during the winter months *[Kolodeznyi and Arshinov, 1970]*. For the Messoyakha field, formulas have been devised to determine the amount of hydrate inhibitors necessary for specific reservoir conditions *[Chersky et al. 1972]*. *Chersky et al. [1968]* have suggested the use of diesel fuel or aerated petroleum fluids with which the gas does not form hydrates. *Parent [1948]* has noted that alcohols might be used to decompose hydrates with respect to the storage of natural gas as a hydrate and its subsequent regasification. The work of *Kobayashi et al. [1951]* on the effect of brine and ethanol solutions on the formation of gas hydrates, suggests that saline formation waters might be cheaper and equally as effective as calcium chloride solutions in decomposing gas hydrates. It is clear that the technology for the optimum recovery of gas reserves in the form of natural gas hydrates is in its infancy, and more research is required so that when reserves are located in North America they may be recovered expeditiously.

CONCLUSIONS

1) Natural gas hydrates exist in nature within narrow temperature and pressure limits in sedimentary basins with relatively thick permafrost sections.

2) The only well documented example is the Messoyakha field in the northern part of the West Siberian basin, which contains 14 trillion ft^3 of ultimate recoverable natural gas, partly in hydrate form. That portion of the field containing natural gas hydrates has 54% higher reserves than would be expected, on the assumption that the reservoir rocks were filled with free gas.

3) Other large gas fields in the northern part of the West Siberian basin and in the Vilyuy basin are believed to contain part of their natural gas reserves in hydrate form. Conclusive evidence of natural gas hydrates in wells drilled in the arctic regions of North America is lacking, though it is anticipated that they will be found.

4) There is great need for careful monitoring of wells drilled in permafrost regions because the nature of natural gas hydrates makes them difficult to detect unless the proper precautions are taken.

5) The technology for drilling in natural gas hydrate zones and for the optimum recovery of gas reserves in the form of natural gas hydrates is in its infancy, and more research is required.

ACKNOWLEDGMENTS

Dr. Alan Judge (Seismology Division, Earth Physics Branch, Department of Energy, Mines and Resources, Ottawa) supplied the determinations of permafrost thickness from northern Canada and the writer is most grateful for this information. This paper has benefited from the critical reviews of Dr. D.B. Robinson (Department of Chemical and Petroleum Engineering, University of Alberta, Edmonton, Alberta), Dr. R.J.E. Brown (Division of Building Research, National Research Council, Ottawa, Ontario), Professor S.L. Miller (Department of Chemistry, University of California at San Diego, San Diego, California), Professor I.R. Kaplan (Department of Geology, University of California at Los Angeles, Los Angeles, California), and Dr. H.W. Habgood, Chief, Physical Sciences Branch and Dr. R. Green, Chief, Earth Sciences Branch, both of Alberta Research, and the writer thanks all these reviewers for their comments.

Contribution No. 623 from Research Council of Alberta.

REFERENCES

Adams, W. D., Developments in Alaska in 1971, *Amer. Ass. Petrol. Geol. Bull.*, *56* (7), 1175, 1972.

Allen, K. W., and G. A. Jeffrey, On the structure of bromine hydrate, *J. Chem. Phys.*, *38*, 2304, 1963.

Anonymous, Earth sciences studies in Arctic marine waters, 1970, *Ocean Science Reviews 1969/70*, *11-24*, Bedford Institute, Dartmouth, Nova Scotia, 1971.

APOA, Chance of gas in Arctic permafrost makes well control equipment vital (Arctic Petroleum Operators Association drilling guide, No. 16), *Oilweek*, *23* (41), 12, 1972.

Baranov, I. Ya., Geographical distribution of seasonally frozen ground and permafrost, in *Principles of Geocryology*, part 1, ch. 7, Akad. Nauk SSSR, Moscow, USSR, 1959. *Nat. Res. Coun. of Can. Tech.*, Transl. 1121, 1964.

Bhatnagar, V. M., Clathrate compounds of water, *Research*, *15* (7) 299, 1962.

Brewer, M. C., Some results of geothermal investigations of permafrost in northern Alaska, *Trans. Amer. Geophys. Union*, *39* (1), 19, 1958.

Brown, J. F., Inclusion compounds, *Sci. Amer. 207* (1), 82, 1962.

Brown, R. J. E., Permafrost in Canada, *Geol. Surv. Can.*, Map 1246A, 1967a.

Brown, R. J. E., Comparison of permafrost conditions in Canada and the USSR, *The Polar Record*, *13* (87), 741, 1967b.

Brown, R. J. E., Permafrost in Canada, Univ. of Toronto Press, 1970.

Brown, R. J. E., Permafrost in the Canadian Arctic Archipelago, *Z. Geomorphol.*, *13*, 102, 1972.

Byk, S. Sh., and V. I. Fomina, Gas hydrates, *Russian Chem. Rev.*, *37* (6), 469, 1968.

Chersky, N., and Yu. Makogon, Solid gas--world reserves are enormous, *Oil and Gas Int.*, *10* (8), 82, 1970.

Chersky, N. V., Yu. F. Makogon, and D. I. Medovsky, Hydrate formation during exploration, development, and experimental production of natural gas deposits in northern areas of the U.S.S.R., *Geol. Str. Neftegazonos. Vost. Chasti Sib. Platformy Prilegayushchikh Raionov, Mater. Vses. Soveshch. Otsenke Neftegazonos. Territ. Yakutii*, *1966*, pp. 458-468, Nedra, Moscow, U.S.S.R., 1968.

Chersky, N. V., V. P. Tsarev, A. V. Bubnov, I. D. Efremov, and E. A. Bondarev, Methods of locating, opening up, and exploiting productive horizons containing crystal hydrates of natural gas (with the example of the Messoyakha field), in *Physical and Engineering Problems of the North*, pp. 112-119, Institute of Physical and Engineering Problems of the North, Academy of Sciences of the U.S.S.R., Siberian Division, Yakut Branch, Nauka, Novosibirsk, U.S.S.R., 1972.

Claussen, W. F., Suggested structures of water in inert gas hydrates, *J. Chem. Phys.*, *19*, 259 and 662, 1951a.

Claussen, W. F., A second water structure for inert gas hydrates, *J. Chem. Phys.*, *19*, 1425, 1951b.

Davidson, D. W., The motion of guest molecules in clathrate hydrates, *Can. J. Chem.*, *49*, 1224, 1971.

Eardley, A. J., Arctic oil and gas reserves: a preliminary estimate, *Twenty-fourth Int. Geol. Congr. (Montreal)*, Sect. 5, pp. 125-134, 1972.

Eremenko, N. A., G. P. Ovanesov, and V. V. Semenovich, Status of oil and gas prospecting in USSR in 1971, *Amer. Ass. Petrol. Geol. Bull.*, *56* (9), 1711, 1972.

Ferrians, O. J., Permafrost map of Alaska, *U. S. Geol. Surv., Misc. Geol. Invest., Map* I-445, 1965.

Halbouty, M. T., A. A. Meyerhoff, R. E. King, R. H. Dott, H. D. Klemme, and T. Shabad, World's giant oil and gas fields, geologic factors affecting their formation, and basin classification. Part I, Giant oil and gas fields, in *Geology of Giant Petroleum Fields*, edited by M. T. Halbouty, pp. 502-528, Mem. 14, Amer. Ass. Petrol. Geol., Tulsa, Oklahoma, 1970.

Hitchon, B., Fluid flow in the western Canada sedimentary basin. 1. Effect of topography, *Water Resour. Res.*, *5* (1), 186, 1969a.

Hitchon, B., Fluid flow in the western Canada sedimentary basin. 2. Effect of geology, *Water Resour. Res.*, *5* (2), 460, 1969b.

Howitt, F., Permafrost geology at Prudhoe Bay, *World Petrol. 42* (8), 28 and 37, 1971.

Jeffrey, G. A., Water structure in organic hydrates, *Accounts of Chem. Res.*, *2*, 344, 1969.

Jeffrey, G. A., Pentagonal dodecahedral water structure in crystalline hydrates, *Materials Res. Bull.*, *7*, 1259, 1972.

Jeffrey, G. A., and R. K. McMullan, The structures of the gas hydrates. Some possible new phases (abstract), *144th Amer. Chem. Soc. Spring Meeting*, 17R-18R, 1963.

Jhaveri, J., and D. B. Robinson, Hydrates in the methane-nitrogen system, *Can. J. Chem. Eng.*, *43*, 75, 1965.

Johnston, G. H., and R. J. E. Brown, Some observations on permafrost distribution at a lake in the Mackenzie Delta, N.W.T., Canada, *Arctic*, *17* (3), 162, 1964.

Katz, D. L., Prediction of conditions for hydrate formation in natural gases, *Trans. AIME.*, *160*, 140, 1945.

Katz, D. L., Depths to which frozen gas fields (gas hydrates) may be expected, *J. Petrol. Tech.*, *23* (4), 419, 1971.

Katz, D. L., Depths to which frozen gas fields may be expected-- Footnotes, *J. Petrol. Tech.*, *24* (5), 557, 1972.

Katz, D. L., D. Cornell, R. Kobayashi, F. H. Poettmann, J. A. Vary, J. R. Elenbaas, and C. F. Weinaug, *Handbook of Natural Gas Engineering*, Chapter 5, Water-Hydrocarbon Systems, pp. 189-221, McGraw-Hill Book Company, Inc., New York, 1959.

Kobayashi, R., H. J. Withrow, G. B. Williams, and D. L. Katz, Gas hydrate formation with brine and ethanol solutions, *Proc. Thirteenth Ann. Convention, Natural Gasoline Ass. Amer.*, 27, 1951.

Kolodeznyi, P. A., and S. A. Arshinov, The technology of pumping antihydrate inhibitor into wells of the Messoyakha Field, *Nauch.-Tekh. Sb. Ser. Gazovoe Delo*, *11*, 9, 1970.

Lachenbruch, A. H., Thermal effects of the ocean on permafrost, *Geol. Soc. Amer. Bull.*, *68* (11), 1515, 1957.

Lachenbruch, A. H., G. W. Greene, and B. V. Marshall, Permafrost and the geothermal regimes, in *Environment of the Cape Thompson Region, Alaska, 1966*, pp. 149-169, USAEC Div. Tech. Inf., 1966.

Mackay, J. R., Gas-domed mounds in permafrost, Kendall Island, N.W.T., *Geog. Bull.*, *7* (2), 105, 1965.

Mackay, J. R., The world of underground ice, *Ann. Ass. Amer. Geograph.*, *62* (1), 1, 1972a.

Mackay, J. R., Offshore permafrost and ground ice, southern Beaufort Sea, Canada, *Can. J. Earth Sci.*, *9* (11), 1550, 1972b.

Makogon, Yu. F., Natural gases in the ocean and the problem of their hydrates, *Ekspress-informatsiya*, *11*, 11, 1972.

Makogon, Yu. F., and D. I. Medovskiy, The possibility of hydrate formation in natural gas deposits on Soviet territories, *Minist. Gazov. Prom., Geolog.*, *1*, 52, 1969.

Makogon, Yu. F., F. A. Trebin, A. A. Trofimuk, V. P. Tsarev, and N. V. Chersky, Detection of a pool of natural gas in a solid (hydrated gas) state, *Dokl. Akad. Nauk SSSR*, *196* (1), 203, 1971.

Makogon, Yu. F., V. P. Tsarev, and N. V. Chersky, Formation of large natural gas fields in zones of permanently low temperatures, *Dokl. Akad. Nauk SSSR, 205* (3), 700, 1972.

Miller, D. J., T. G. Payne, and G. Gryc, Geology of possible petroleum provinces in Alaska, *U. S. Geol. Surv. Bull.,* 1094, 1959.

Noaker, L. J., and D. L. Katz, Gas hydrates of hydrogen sulfide-methane mixtures, *Trans. AIME., 201,* 237, 1954.

Parent, J. D., The storage of natural gas as hydrate, *Inst. Gas Tech. Res. Bull., 1,* 1948.

Pauling, L., and R. E. Marsh, The structure of chlorine hydrate, *Proc. Nat. Acad. Sci., 38,* 112, 1952.

Robinson, D. B., and J. M. Hutton, Hydrate formation in systems containing methane, hydrogen sulphide and carbon dioxide, *J. Can. Petrol. Tech., 6* (1), 6, 1967.

Schröder, W., Review of gas hydrates, *Samml. Chem. und Chem.-Tech. Vortr., 29,* 1, 1927.

Sokolov, V. A., Gas geochemistry in the Earth's crust and atmosphere, Nedra, Moscow, U.S.S.R., 1966.

Stackelberg, M. von, Solid gas hydrates, *Naturwissenschaften, 36,* 327 and 359, 1949.

Stackelberg, M. von, and H. R. Müller, On the structure of gas hydrates, *J. Chem. Phys., 19,* 1319, 1951.

Stoll, R. D., Effects of gas hydrates in sediments, in *Natural Gases in Marine Sediments,* edited by I. R. Kaplan, pp. 235-248, Plenum Press, New York, 1974.

Stoll, R. D., J. Ewing, and G. M. Bryan, Anomalous wave velocities in sediments containing gas hydrates, *J. Geophys. Res., 76* (8), 2090, 1971.

Trebin, F. A., V. A. Khoroshilov, and A. V. Demchenko, Kinetics of hydrate formation of natural gases, *Gazov. Prom., 11* (6), 10, 1966.

Tricart, J., *Geomorphology in Cold Environments,* MacMillan and Co. Ltd., 1970.

Trofimuk, A. A., N. M. Chersky, and V. P. Tsarev, Accumulation of natural gases in hydrate formation zones of the world's oceans, *Dokl. Akad. Nauk SSSR, 212,* 931, 1973.

Unruh, C. H., and D. L. Katz, Gas hydrates of carbon dioxide-methane mixtures, *Trans. AIME., 186,* 83, 1949.

Vasiliev, V. G., Yu. F. Makogon, A. A. Trofimuk, and N. V. Cherskiy, Deposits of solid gas, Preprint, 12th World Gas Conf., Nice, 1973.

Wilcox, W. I., D. B. Carson, and D. L. Katz, Natural gas hydrates, *Ind. Eng. Chem., 33* (5), 662, 1941.

EXPERIMENTS ON HYDROCARBON GAS HYDRATES IN UNCONSOLIDATED SAND

P.E. Baker

Chevron Oil Field Research Company
P.O. Box 446
La Habra, California   90631

ABSTRACT

Experiments were carried out to observe the formation and decomposition of hydrocarbon gas hydrates in an unconsolidated sand pack 4.4 cm in diameter and 45.7 cm long. Pressure and temperature ranges were 37 to 43 bars and 5 to 10°C; gas used was 90% methane and 10% ethane. Different procedures were followed, all of which were successful to some extent in making hydrate. Of the saturation range studied, lower initial water content generally resulted in more hydrate. Decomposition and gas production were carried out at constant pressure or constant temperature. In the constant pressure (rising temperature) method, decomposition started when the sand pack temperature reached the hydrate formation temperature for methane, and continued until ethane hydrate formation temperature was reached.

INTRODUCTION

Very large natural gas reserves may exist in the Arctic in hydrate form. The pressure-temperature relation under the permafrost satisfies hydrate existence conditions over a considerable depth interval [*Katz, 1971*]. Several Russian publications have reported the apparent discovery of gas hydrate reserves and production of gas from them [*Chersky, 1970; Makogon, 1971; Anonymous, 1971*]. This type of reserve could conceivably be very attractive because of the large amount of gas that can be concentrated in a hydrate (roughly 160 standard liters of methane per liter of hydrate). On the other hand, such a reserve poses severe production problems because of the heat required to decompose the hydrate, as well as its tendency to

block gas flow. (Only one report has appeared in petroleum literature of a study of gas hydrates in a porous medium [*Evrenos, 1971*].
In that study, the idea was to use hydrates to render a formation
impermeable.) Even after decomposition of the gas hydrate, the
presence of high water saturation will impede production.

There may be an excess of gas or an excess of water in a reservoir containing hydrates, or the two may be in stoichiometric
proportion in some regions. To be economically producible, excess
gas is probably required. During hydrate formation, an excess of
one phase or the other would seem necessary to permit the fluid
movement necessary to maintain local pressure.

Laboratory experiments were carried out to observe the formation
of hydrocarbon gas hydrates in a porous medium (unconsolidated sand).
The nature of these experiments was essentially exploratory in that
the primary purpose was to find out what kind of information can be
obtained from such small scale experiments, and to see what experimental difficulties and techniques would be involved. The study did
provide information on the effect of different initial water contents. Methods and results are reported below.

EXPERIMENTAL METHOD AND APPARATUS

The equipment consists of a sand pack (unconsolidated) in a
pressure vessel, a refrigerated water bath, metered and pressure
regulated sources of water and gas, a back pressure regulator on
the effluent side, a gas (wet test) meter, and an open graduate for
water effluent measurement. Figure 1 shows a schematic diagram of
the apparatus, and Figure 2 shows details of the sand pack. The
sand used was 20 to 30 mesh (Tyler screen, approximately .5 to .7 mm
openings). Permeability was 20 darcies, and porosity 38%.

The gas used was 90% methane and 10% ethane (gravity 0.6 on the
air = 1.0 scale). Gas of approximately this composition frequently
occurs in gas fields; it forms hydrates at relatively low pressures
compared to methane, as shown in Figure 3, following *Katz* [*1959*]
and *Kobayashi and Katz* [*1949*].

Several procedures were used for making the hydrate. One successful procedure was that of *Evrenos* [*1971*] in which water and gas
alternately flow through the core until it is plugged, presumably by
the hydrate. This method is based on the fact that some mechanical
mixing helps promote hydrate formation. The disadvantage is that
it does not permit control of water content or saturation (percent
of pore volume). A preferred method was to establish the desired
water saturation, and leave the core attached to a pressure-regulated
gas supply overnight or longer. When hydrate begins to form, pressure drops and the process tends to stop. To prevent this from

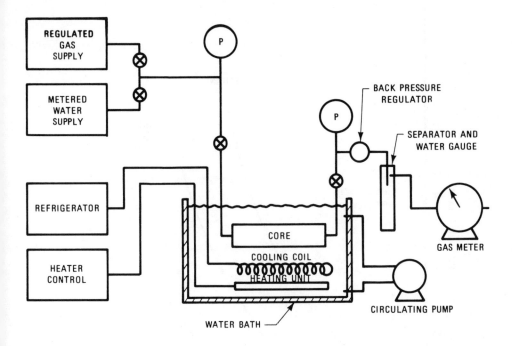

Fig. 1. Schematic of apparatus for making gas hydrate in sand pack.

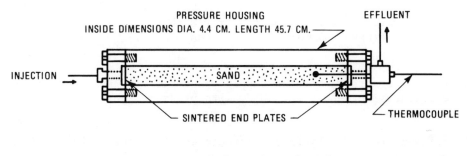

Fig. 2. Details of sand pack.

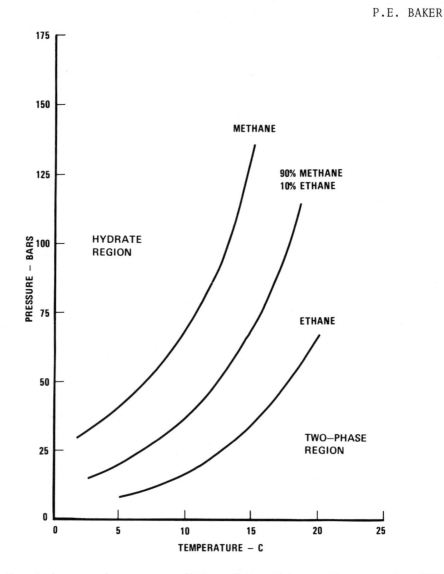

Fig. 3. Hydrate existence condition for methane, ethane, and a 90% methane - 10% ethane mixture. (References 6 and 7).

occurring, a constant pressure supply was employed. Runs were made at temperatures between 5 and 10°C, with pressures between 35 and 43 bars.

The usual test for hydrate existence was to produce the gas from the core (sand) and compare the amount produced with the amount that could have been contained in the core without hydrate

occurrence. Two methods were used to decompose the hydrate and
produce the gas. In one method, the outlet pressure was dropped
while bath temperature was held constant. In the other method, the
bath temperature was gradually raised, while gas was allowed to
escape through a back pressure regulator that maintained constant
core pressure; when all of the hydrate gas appeared to have been
produced, the remaining (free) gas was produced by dropping the
pressure.

The first method did not indicate the onset or completion of
hydrate decomposition. Apparently the rate of heat flow into the
small diameter core was so rapid that it did not limit the decomposition rate, and the production rate was controlled only by the
pressure drop established by means of the back pressure regulator.

RESULTS

The results of ten runs are shown in Table 1. Runs 7 and 8
were performed to test and demonstrate experimental procedures.
The method used in Run 1 was to establish water saturation, with a
gradual increase in pressure with fixed temperature. The apparent
amount of hydrate was very small. Possibly hydrate formation occurred in such a way as to prevent further gas-water contact, while
still permitting gas to reach the down stream pressure gauge. In
the other runs a larger amount of hydrate was formed. But only in
Runs 9 and 10 did it approach the maximum amount permitted by the
water content.

In Runs 9 and 10, although less water was present than in
other runs, the amount of gas hydrated was larger. The reason for
this is that gas contacts the water only in that portion of the
sand where water saturation is somewhat less than 100%. Above some
minimum of water content, additional water accumulates at the bottom of the core, and thus actually reduces the water available for
making hydrate.

In Run 10, the hydrate was broken by increasing temperature,
while holding core pressure constant at 42.4 bars, by means of the
back pressure regulator. Results are shown in Figure 4. As bath
temperature increased, core temperature also increased until it
reached 6.7°C, which is approximately the hydrate formation temperature for methane at 42.4 bars. Onset of gas production and level-
ing off of the core temperature curve indicate hydrate decomposition.
Core and bath temperature finally coincide again at about 16.7°C,
which is the hydrate formation temperature for ethane at the core
pressure. Gas production at this temperature essentially stopped,
with further slow production resulting from thermal expansion.

TABLE 1. Results of Experiments on Gas Hydrates in a Sand Pack

Run Number:	1	2	3	4	5	6	7	8	9	10
Initial Water, cc.	161	162	145	208	178	183	201	203	68	83
Initial Gas Space, cc.	137	136	153	90	120	115	97	95	230	215
Pressure, Bars	21.4–41.8	40.5	42.9	39.9	37.0	38.3	35.7	35.9	35.7	42.4
Temperature, °C	10.0	5.6	6.7	5.0	5.0	5.0	23.9	20.0	4.4	5.6
Gas Produced, S.L.*	6.37	8.78	11.89	7.39	8.13	9.69	3.40	3.71	20.22	24.53
Maximum Free Gas, S.L.	5.60	5.49	6.51	3.65	4.45	4.42	3.26	3.23	8.41	9.09
Maximum Gas in Solution, S.L.†	.15	.17	.14	.24	.20	.20	.20	.20	.07	.07
Gas in Hydrate, S.L.	.59	3.11	5.24	3.74	3.51	5.07	0	0	11.72	15.35
Maximum Possible Hydrate Gas, S.L.†	35.72	35.72	32.00	46.16	39.36	40.50	0	0	15.01	18.32
Hydrate Formation Method	Pressure Build-up 1 hr.	Slug Flow	Pressure Maintenance 64 hr.	Pressure Maintenance 47 hr.	Pressure Maintenance 54 hr.	Pressure Maintenance 24 hr.	Pressure Build-up 5 min.	Pressure Maintenance 18 hr.	Pressure Maintenance 66 hr.	Pressure Maintenance 48 hr.
Production Method	Pressure Reduction	Temperature Increase	Pressure Reduction	Pressure Reduction	Pressure Reduction	Temperature Increase	Pressure Reduction†	Pressure Reduction	Temperature Increase	Temperature Increase

Core properties: 20–30 mesh sand, porosity 38%, permeability 20d, pore volume 293 cc.

*S.L. = standard liters volume.

†Based on water content.

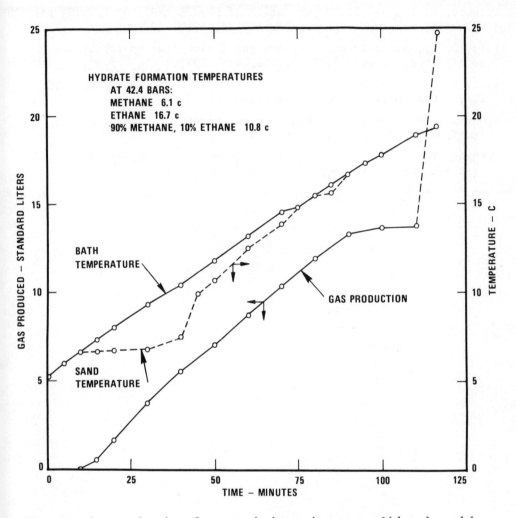

Fig. 4. Gas production from gas hydrate in unconsolidated sand by temperature increase at constant pressure.

REFERENCES

(Anonymous), Solid Gas, *Review of Sino-Soviet Oil, No. 10,* p. 12, 1971. (Excerpted from Sotsialisticheskaya Industriya, p. 2, March 30, 1971.)

Chersky, N., Solid gas - world reserves are enormous, *Oil and Gas International, 10,* 82, 1970.

Evrenos, A. I., *et al.*, Impermeation of porous media by forming gas hydrates *in situ*, *J. Petrol. Tech.*, *23*, 1059, 1971.

Katz, D. L., Depth to which frozen gas fields (gas hydrates) may be expected, *J. Petrol. Tech.*, *23*, 419, 1971.

Katz, D. L., *et al.*, *Handbook of Natural Gas Engineering*, pp. 207-221, McGraw Hill Book Co., New York, 1959.

Kobayashi, R., and D. L. Katz, Methane hydrate at high pressure, *Trans. AIME*, pp. 66-70, March, 1949.

Makogon, Yu. F., *et al.*, Detection of natural gas deposits in solid (gas hydrate) state, *Dokl. Akad. Nauk SSSR*, *196*, 203, 1971.

EFFECTS OF GAS HYDRATES IN SEDIMENTS

Robert D. Stoll

School of Engineering and Applied Science
Columbia University
New York, New York 10027

ABSTRACT

Natural gas and water combine to form an ice-like substance called a gas hydrate under certain conditions of temperature and pressure. The presence of large quantities of gas in some sediments recovered during deep sea coring operations suggests that gas hydrates may exist in some of the sea floor sediments, because environmental conditions over vast areas of the ocean bottom are conducive to hydrate formation. Experiments on sediments containing artificially formed hydrate have shown that acoustic wave velocity and other physical properties may be markedly changed because of the presence of the hydrate. A program of research is currently being carried out to determine the acoustic, thermal and mechanical properties of sedimentary materials containing gas hydrates under conditions similar to those in the marine environment.

INTRODUCTION

Ice-like substances called gas hydrate form at temperatures well above freezing when certain gases are combined with water at elevated pressures. The structure of hydrates is such that gas molecules are trapped within a framework of hydrogen bonded water molecules, producing a stable crystalline structure. The list of gases known to form hydrates includes: methane, ethane, propane, hydrogen sulfide, carbon dioxide, nitrogen, and many others of small molecular size as well as many naturally occurring gases that are mixtures of these components. Since the identification of chlorine hydrate by Davy in 1810, there have been many studies of gas hydrates pertaining to equilibrium thermodynamics and crystalline structure,

235

as well as to various applications in science and industry. Extensive bibliographies relating to this work are contained in review articles by *van der Waals and Platteeuw* [*1959*], *Jeffrey and McMullan* [*1967*], *Deaton and Frost* [*1946*] and *Katz et al.* [*1959*].

The existence of gas hydrates in the natural environment of the earth and the planets has become a topic of increasing interest over the past two decades. There is evidence that air hydrates may be trapped in the polar ice caps [*Miller, 1969*], and that gas hydrates may exist on several of the planets [*Miller, 1961*]. Formations of natural gas hydrates have been found in sediments under the permafrost in Siberia [*Chersky and Makogon, 1970*]. Russian scientists have claimed that the extensive deposits in Siberia are being utilized as a new source of natural gas.

The possibility that gas hydrates exist in a marine environment was suggested at Lamont-Doherty Geological Observatory in 1970 [*Stoll et al. 1971*] after reviewing reports of unusual acoustic records that seemed to be related to the presence of very large amounts of gas in the sediment [*Ewing and Hollister, 1972*]. An acoustic reflector was observed to have the same profile as the sea floor, but cut across clearly defined bedding planes, and appeared to be related to the presence of the gas. An example of this kind of record from the Blake-Bahama outer ridge is shown in Figure 1. In addition to the impedance change indicated by this unusual reflector, wave velocities, based on the most obvious correlation between reflections and lithologic changes, were found to be over 2 km/sec, which is anomalously high when compared to velocities in sediments with similar porosity and grain size that contain no gas. Recent Sonobouy measurements have confirmed the high velocities originally recorded [*Bryan, 1974*]. In the past two years, many similar unexplained acoustic records have been reported in areas containing terrigeneous material. This suggests that gas hydrates may exist or have influenced the development of sediment profiles at numerous sites near the continental margins of the earth.

At the time of the initial discovery, a research program was initiated at Lamont-Doherty Geological Observatory to investigate the extent to which hydrates do exist in nature, and the effect they have on the physical properties of marine sediments. This program has included extensive experimental work on sediments containing hydrate formed under laboratory conditions, as well as field work designed to obtain samples of naturally occurring hydrates. The field work was undertaken on Leg 19 of the Deep Sea Drilling Project in an area of the Bering Sea where acoustic records and the character of the sediment were both considered favorable for the existence of hydrates [see *Bryan, 1974*]. All of our laboratory and field work to date strongly support the hypothesis that gas hydrates do exist at various locations in the ocean sediments, and that they may have a marked effect on acoustic, thermal, and other physical properties.

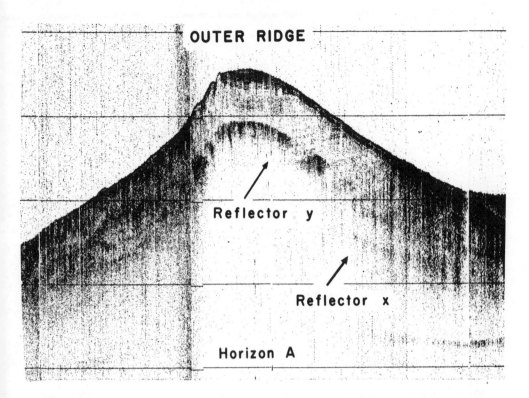

OUTER RIDGE

Reflector y

Reflector x

Horizon A

Fig. 1. Record of acoustic sounding on the Blake-Bahama outer ridge from *Markl et al.* [*1970*].

Some idea of the extent to which the marine environment is favorable to hydrate formation can be derived from the extensive data on natural gas hydrates already published (see *Deaton and Frost* [*1946*] and *Katz et al.* [*1959*]). Most of the work on natural gas hydrates was stimulated by a problem encountered by the gas industry resulting from the formation of hydrates in damp gas and the clogging of high pressure pipelines. The temperature and pressure relationship for hydrate formation for methane and for two natural methane-rich gases are shown in Figure 2. This information, together with a reasonable estimate of the thermal gradient, can be used to make a rough calculation of the depth in the sediment to which temperature and pressure conditions would sustain hydrates. As an example, curve CC in Figure 2 was extrapolated to cover the desired range of pressure, the curve for pure methane being used as a guide. After correcting for the effect of the salinity of sea water, the maximum potential depth of hydrate formation was calculated for different water depths and bottom temperature as shown in

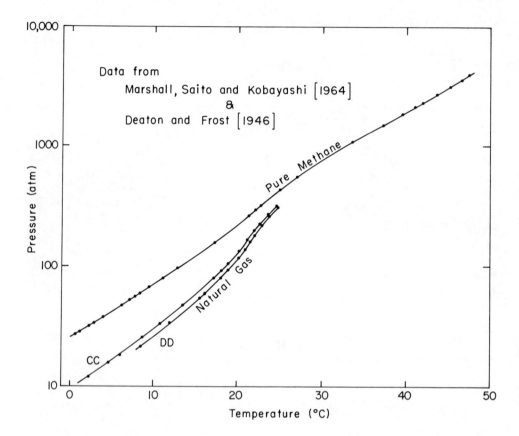

Fig. 2. Hydrate equilibrium conditions for pure methane and natural gas.

Figure 3. In these calculations the density of the water was assumed to be constant, and the thermal gradient was chosen as .035°C per meter. This is somewhat less than the average value found in many gas-free sediments, and was chosen to account for an assumed increase in thermal conductivity in sediment containing hydrate. The actual gradient could vary over a fairly wide range, depending on the hydrate concentration, sediment thickness and other variables.

The results shown in Figure 3 suggest that the bottom environment over a very large per cent of the sea floor is conducive to the formation of hydrate given a sufficient amount of gas to saturate the interstitial water. The thickness of potential zones of formation in the sediment range from zero in relatively shallow water to over a kilometer in the deep oceans. Within these zones, amounts

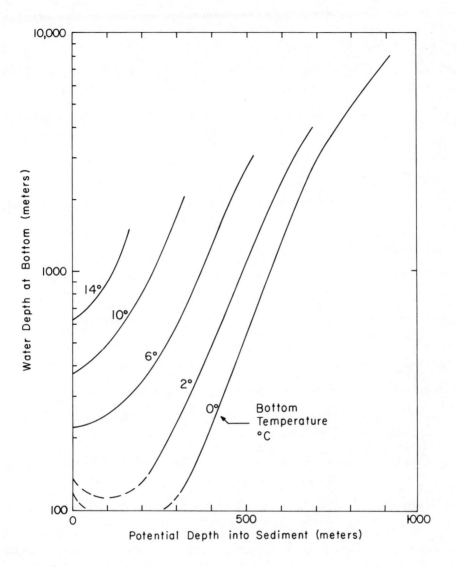

Fig. 3. Potential depth of hydrate formation in sediments from natural gas. The hydrate is stable under conditions described by depth of water and sediment to the left of each line.

of hydrate could vary widely, depending on the number of gas molecules available at a given depth. Thus, very complex patterns of physical properties may exist that differ from one location to another.

If one considers the effects of normal freezing on the physical properties of sediments, it is reasonable to expect that somewhat similar effects occur in deep-sea sediments containing hydrates. Marked changes in elastic constants, thermal conductivity, and the ultimate shearing strength of the sediments are to be expected if the concentration of hydrate is high. In turn, these properties exert a controlling influence on acoustic reflectivity, velocity profiles, heat flow, and the structure of the sea bottom in general. Thus, the presence of gas hydrates may have a significant influence in many areas of technology, including interpretation of acoustic sounding and bottom-bounce sonar records, exploration and drilling for offshore petroleum and the design of ocean bottom structures.

PHYSICAL PROPERTIES

As a first step in studying the effects of gas hydrates on sediments, a series of experiments are being performed to measure acoustic wave velocities, thermal conductivity and other physical properties in sediment containing various hydrates (methane, propane, ethane, etc.). The experimental setup shown in Figure 4 is a modified version of equipment used in preliminary experiments to investigate acoustic wave velocities [see *Stoll et al. 1971*]. In most of these experiments, a pressure vessel containing water-saturated sediment was placed in a cooling bath, and commercial grade methane gas was used to pressurize the system and to supply gas for hydration. For the preliminary experiments, a uniform standard testing sand with rounded grains was used (Ottawa sand passing U.S. Std. #20 sieve and retained on the #30 sieve). Sediment saturated with sea water was confined between two porous plates and gas was bubbled upwards through the specimen, with the rate controlled by a needle valve at the top of the chamber. A piston through the top of the pressure vessel allowed the intergranular stress in the specimen to be controlled to simulate varying overburden pressures. As the force on the piston was measured continuously, the piston also proved to be a valuable check on the gas-water pressure at the top of the chamber. Waves were generated and received by transducers utilizing lead zirconate-titanate disks polarized to respond in the thickness mode. The sending element was pulsed 100 times per second, and the time to the first arrival of a signal at the receiving transducer was measured by means of a calibrated delay circuit.

The experiments were performed at temperatures in the range of 2 to 4°C and with gas pressures varying from 700 to 1200 psia. Gas was bubbled through the specimen, as the pressure was slowly cycled through the range of 1200 to 700 psia many times. As hydrate accumulated in the sediment, the wave velocity was found to increase from 1.85 km/sec to over 2.7 km/sec. At the conclusion of some of the experiments, the test chamber was rapidly dismantled and the

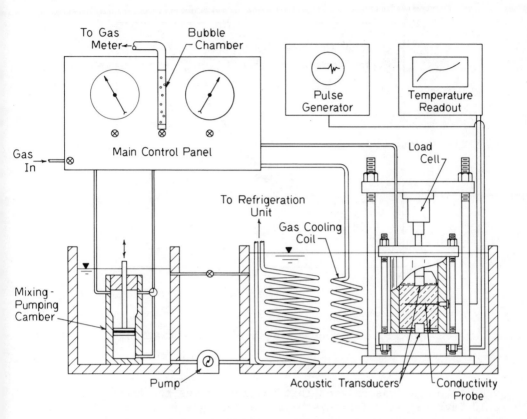

Fig. 4.   Test setup.

specimen was found to be hard and coherent, closely resembling frozen, saturated sand. In these cases, the dissociation of gas and water required considerable time (typically, over 30 minutes) at room temperature and pressure.

The heat absorbed by dissociation of gas hydrate produces a temperature stabilizing effect that may be observed experimentally. This technique has been used by many investigators to establish the equilibrium diagrams for different kinds of hydrate. As an example, the temperature-pressure history during dissociation at the end of one of our experiments is shown in Figure 5. In this case, the test chamber was allowed to warm slowly from the test temperature to room temperature while the chamber pressure was held constant at 1000 psi. Before warming, velocity increase, a restriction of gas flow, and all the signs of hydrate formation had been observed. The plateau of almost constant temperature between 8 and 9°C is a clear indicator of dissociation, and over the corresponding time interval a

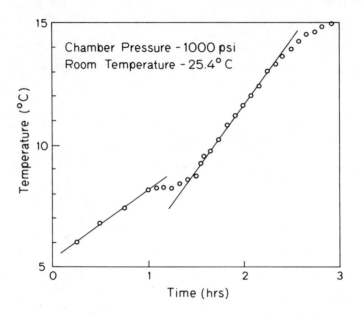

Fig. 5.   Warming curve showing dissociation.

quantity of gas was emitted that was many times the amount that would have resulted from dissolved or free gas in the test chamber. Despite the position of the thermistor probe and the complex temperature gradients that undoubtedly exist in the specimen, the temperature and pressure at dissociation agree favorably with the predicted equilibrium curve for methane and sea water.

The measured velocity of compressional waves for dry and saturated sediment and for sediment containing hydrate is shown in Figure 6. Data from *Hardin and Richart* [*1963*] and from *Whitman and Lawrence* [*1963*] for compressional and shear wave velocity in the same standard sand is also included in Figure 6 for comparison. The wave velocity in sediments containing naturally occurring hydrate may range well above and below the values recorded in Figure 6, because these values depend on the particular set of conditions in our experiments. In view of the fact that a continuous flow of gas through the specimen was used to produce interstitial hydrate, it is likely that the quantity of hydrate was somewhat below the maximum possible, because of clogging and isolation of gas filled void spaces in the final stages of the experiments. A restricted rate of gas flow and often complete blockage, which resulted in a differential pressure build-up across the specimen, was observed in most tests. For this reason, careful measurements of sediment porosity and the quantities of hydrate, water and gas are being made

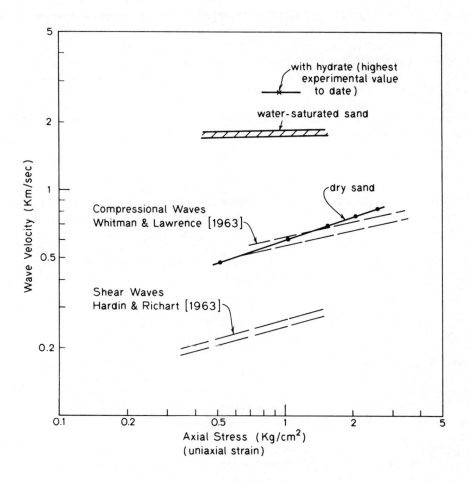

Fig. 6. Wave velocity in sediment for various conditions

in current experiments to provide the data that must be considered for interpretation of the measured thermal and mechanical properties.

Factors that may influence the thermal and mechanical properties include temperature, the quantity of hydrate compared to water and/or gas in the interstices of the sediment, and the bond between the hydrate and the sediment grains. Relatively small amounts of crystalline hydrate adhering to the sediment particles at their contact points or faces may increase the stiffness of the sediment skeleton in much the same manner as cementation or lithification. This could produce marked changes in wave velocity, thermal conductivity, and other sediment properties, even though the quantity

of gas was well below the amount required to combine with all of
the available water in the interstices of the sediment. Another
possibility is that temperature effects, such as those observed in
ice, will also be found in hydrates. Ice becomes less compressible,
and its shear modulus increases at lower temperatures as reflected
by the higher wave velocities that are observed. The velocity of
compressional waves in solid sea ice increases from 2.8 km/sec at
-2°C to 3.5 km/sec at -15°C according to *Untersteiner* [*1966*]. Shear
wave velocity increases from 1.55 to 1.85 km/sec over the same temperature range.

Another factor which is particularly important in experimental
work is the influence of free gas in the water and/or hydrate filling
the interstices of the sediment. Acoustic wave velocity in water
filled sediments is known to be markedly affected by the presence
of free gas [*Brandt, 1960*]. A drastic reduction of velocity caused
by gas bubbles, and the resulting displacement of some of the interstitial water prior to hydrate formation is quite obvious in the
early stages of our current experiments.

Thermal conductivity measurements may also exhibit a strong dependence on the amount of free gas. Some insight into this problem
can be obtained from the results of studies on soil freezing. The
effects of free gas on the thermal conductivity of frozen and unfrozen granular sediment is shown in Figure 7. From this figure,
it is obvious that relatively small amounts of gas can cause significant changes in the thermal conductivity of frozen sediments. It
is likely that the same type of effect will be found in sediments
containing ice-like, crystalline gas hydrate.

Much of our recent laboratory work has been devoted to extensive modification and improvement of the equipment used in preliminary experiments. New acoustic transducers have been built to with-
stand long periods of exposure to gas saturated water under high
pressure. Several new components have been added to the system to
permit better control over the formation of hydrate and the measurement of additional physical properties, principally those involving
heat flow. The mixing-pumping chamber shown in Figure 4, as well
as the modified main test chamber, were first constructed of acrylic
plastic to permit observation of hydrate formation, bubble pattern,
mixing, and other internal phenomena. Because of pressure limitations, propane was substituted for methane for experiments utilizing
the plastic components. For work with methane and natural gas at
high pressure, the plastic has been replaced with high strength aluminum alloy.

In addition to the work described above, new techniques that do
not depend on the extensive and continuous flow of free gas through
the sediment column are being investigated as a means of producing

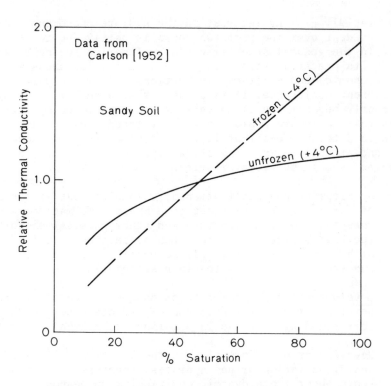

Fig. 7. Thermal conductivity for frozen and unfrozen sand.

hydrate in sediments. One method that is being developed is to mix preformed hydrate, water, and sediment particles, after which the solids are compacted to form a stable mass of appropriate density. This technique eliminates the formation of isolated gas bubbles.

To summarize, all of our experimental data to date indicate that the presence of hydrate can produce significant changes in the physical properties of marine sediments. These changes include modifications of acoustic impedance, thermal conductivity, and compressibility. Moreover, properties of sediments containing appreciable amounts of hydrate appear to be similar in many respects to those of frozen water-filled sediment. With the accumulation of more quantitative data appropriate to the marine environment, we expect to be able to identify and explain the unusual acoustic signatures that recently have been reported in many gas-containing sediments near the continental margins.

RECOMMENDATIONS FOR FUTURE WORK

The rapidly growing interest in the hydrate phenomenon that has been evident over the past two years is partly due to its importance in the general scientific study of marine sediments and partly to its practical significance in a variety of areas with ecological, economic, and military implications. From the participation in the present conference, it is evident that a number of groups have recently begun research on various aspects of the hydrate problem and that there are a wide range of related areas. Based on our work to date, we feel that if this research is to proceed in an orderly and productive way, investigations in three areas, as outlined below, should be given priority:

A. Development and utilization of a sampler that is capable of verifying the existence of a hydrate formation *in situ*. Once the acoustic signature and other physical phenomena (porosity, structure, etc.) can be related to the existence of hydrate *in situ*, it will be possible to locate, identify, and study potential sites in a systematic manner;

B. Development of further data on changes of physical properties owing to the presence of hydrate under conditions corresponding to those in a marine environment;

C. Development of information on the kinetics of formation and dissociation of gas hydrates. Whereas a great deal is known about thermodynamic equilibrium corresponding to phase change, the literature contains little information on the kinetic aspects of the problem. The importance of this information became apparent after it was found that our experimental work was being hindered by a lack of knowledge of factors affecting the kinetics of formation. Future work dealing with the origins of hydrates in sediments and their effect on the development of the sediment profile, strongly depend on a knowledge of the kinetic response.

ACKNOWLEDGMENTS

This paper is based in part on research sponsored by the National Science Foundation, under Grant GA 30881 to Columbia University, and by the Office of Naval Research, under Contract N00014-67-A-0108-0004.

REFERENCES

Brandt, H., Factors affecting compressional wave velocity in unconsolidated marine sand sediments, *J. Acoust. Soc. Am.*, *32*, 171, 1960.

Bryan, G. M., *In situ* indications of gas hydrates, in *Natural Gases in Marine Sediments*, edited by I. R. Kaplan, pp. 299-308, Plenum Press, New York, 1974.

Carlson, H., Calculation of depth of thaw in frozen ground, *Highway Res. Bd., Spec. Rep.*, *2*, 1952.

Chersky, N., and Y. Makogon, Solid gas-world reserves are enormous, *Oil and Gas Int.*, *10*(8), 82, 1970.

Deaton, W. M., and E. M. Frost, Jr., Gas hydrates and their relation to the operation of natural gas pipe lines, *U. S. Bur. of Mines Monogr.* *8*, 1946.

Ewing, J. I., and D. C. Hollister, *Initial Reports of the Deep Sea Drilling Project*, vol. 11, pp. 951-973, U. S. Government Printing Office, Washington, 1972.

Hardin, B. O., and F. E. Richart, Elastic wave velocities in granular soils, *J. Soil Mechs. and Found. Div.*, *ASCE, 89*, (SMI), 33, 1963.

Jeffrey, G. A., and R. K. McMullan, The clathrate hydrates, *Progr. Inorg. Chem.*, *8*, 43, 1967.

Katz, D. L., D. Cornell, R. Kobayashi, F. H. Poettmann, J. A. Vary, J. R. Elenbaas, and C. F. Weinang, *Handbook of Natural Gas Engineering*, P. 802, McGraw-Hill, New York, 1959.

Markl, R. G., G. M. Bryan, and J. I. Ewing, Structure of the Blake-Bahama outer ridge, *J. Geophys. Res.*, *75*, 4539, 1970.

Marshall, D. R., S. Saito, and R. Kobayashi, Hydrates at high pressure: Part 1 - methane-water, argon-water, and nitrogen-water systems, *AIChE J.*, *10*, 202, 1964.

Miller, S. L., The occurrence of gas hydrates in the solar system, *Proc. Nat. Acad. Sci.*, *47*, 1798, 1961.

Miller, S. L., Clathrate hydrates of air in Antarctic ice, *Science*, *165*, 489, 1969.

Stoll, R. D., J. Ewing, and G. M. Bryan, Anomalous wave velocities in sediments containing gas hydrates, *J. Geophys. Res.*, *76*, 2090, 1971.

Untersteiner, N., Sea ice, in *Encyclopedia of Oceanography*, vol. 1, edited by R. W. Fairbridge, pp. 777-781, Reinholt, New York, 1966.

van der Waals, J. H., and J. C. Platteeuw, *Advances in Chemical Physics*, vol. 2, pp. 2-57, Interscience, New York, 1959.

Whitman, R. V., and F. V. Lawrence, Discussion of paper by *Hardin, and Richart* [*1963*], *J. Soil Mechs. and Found. Div.*, *ASCE, 89* (SM5), 112, 1963.

ACOUSTICS AND GAS IN SEDIMENTS: APPLIED RESEARCH LABORATORIES (ARL) EXPERIENCE

Loyd D. Hampton and Aubrey L. Anderson

Applied Research Laboratories
The University of Texas at Austin
Austin, Texas 78712

ABSTRACT

A discussion of the acoustical properties of saturated sediments and of gas bubbles in water is given as background to a discussion of the present state of knowledge of the acoustical properties of gas bearing sediments.

Four Applied Research Laboratories (ARL) programs are discussed, which have the commonality of gas in sediments, and which relate more or less directly to acoustic behavior. The first program is a series of laboratory measurements of the acoustic properties of constructed and controlled sediments. During these measurements, it became obvious that presence of gas dominated the acoustic behavior. The next program includes development of a chemical model of a gassy sediment and the design and construction of a diver operated sampler for collecting uncontaminated gas samples from a gassy sediment.

The third program is the development of a method for measuring sound speed during sediment coring. The cutting head of the coring barrel is modified to include two transducers and sound speed is recorded as the core barrel penetrates the sediment. The fourth program is an expansion of the sound speed measurement to include both attenuation (by amplitude measurement) and acoustic volume scattering (by use of an additional receiver at a right angle to the primary acoustic path). As the scattering mechanism is different for liquid saturated and gas bearing sediments, this technique should also be useful for indicating the presence of gas.

INTRODUCTION

Gas bubbles, when present in sediments in even small amounts, dominate the acoustical characteristics of the sediment. Acoustic attenuation, propagation speed, and sediment reflective character- istics are different for liquid saturated and partially saturated (gas bearing) sediments. Presence of gas bubbles also controls the acoustical propagation in water. The acoustical properties of sat- urated sediments and of bubbly water will be summarized as back- ground to a discussion of the acoustical properties of gas bearing sediments.

A. Acoustical Properties of Saturated Sediments

Attenuation of sound in saturated sediments depends on fre- quency of the propagating sound and on properties of the sediment such as grain size and porosity. Figures 1, 2, and 3 summarize mea- sured attenuation values reported by several investigators for the frequency range from 1 to 1000 kHz. These data are given to indi- cate the order of magnitude of attenuation to be expected for sat- urated sediments of various types and to allow later comparison with attenuation values for gassy sediments.

Figures 1, 2, and 3 include attenuation data from both labora- tory and field measurements of natural and artificial sediments. Figure 1 summarizes available data for clays and silts (no more than 1% material larger than the mud-sand boundary of 62.5 μ or 4 ϕ). The data of Figure 2 are for muddy sands (more than 10% sand size material), while those of Figure 3 are for pure sand (no more than 1% material smaller than 62.5 μ). There is a tendency at all fre- quencies for attenuation to be higher in either sand or muddy sand sediments than in clays and silts. This is most pronounced for the low frequencies.

Only very slight dependence of sound speed in saturated sedi- ments on frequency has been reported [*Hampton, 1967*]. The sound speed is a function of porosity as shown in Figure 4 [*Akal, 1972*]. The normal incidence interface reflection coefficient is strongly correlated with sediment porosity [*Faas, 1969; Hamilton, 1970; Akal, 1972*]. High porosity, small grain size clays and silts are poor reflectors. Acoustic reflectivity increases with increasing sand content.

B. Acoustical Properties of Bubbly Water

Basic to many aspects of the acoustical properties of bubbly water is the fact that gaseous bubbles in water are capable of vi- bratory motion with a sharply peaked resonance at the fundamental

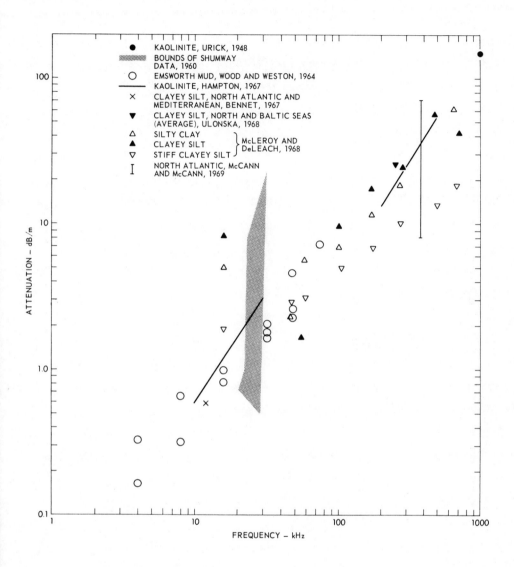

Fig. 1. Acoustical attenuation versus frequency in clays and silts (less than 1% sand).

pulsation frequency. Motion of the water in the vicinity of the bubble is controlled by the internal pressure, which varies inversely with volume; the gas bubble acts as the cavity and the surrounding water as the vibrating mass of an acoustical oscillator. The resonance frequency of a gas bubble in water has been investigated by *Minnaert* [*1933*]*, Richardson* [reported by *Briggs et al. 1947*] and

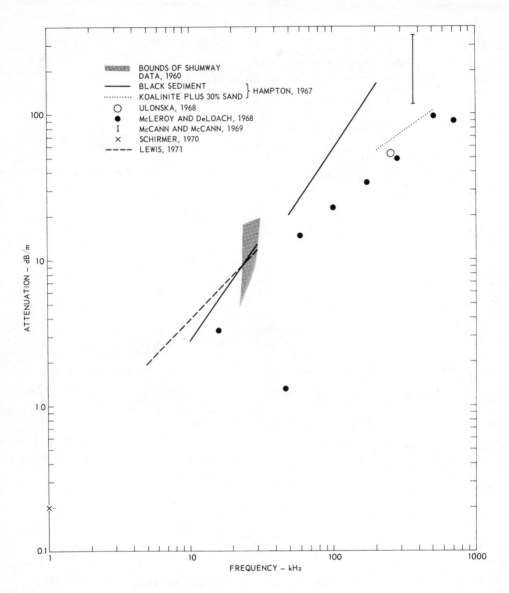

Fig. 2. Acoustical attenuation versus frequency in muddy sand and sandy mud.

Houghton [1963]. A general equation for the fundamental resonance frequency, which includes the earlier ones as special cases, is given by *Shima* [1971].

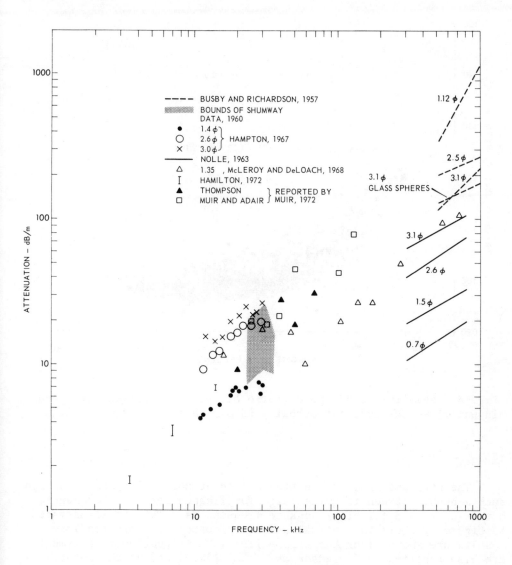

Fig. 3.   Acoustical attenuation versus frequency in sand.

Three distinct frequency ranges are identifiable with different acoustical behavior in each: frequencies below the fundamental resonance, frequencies in the vicinity of the fundamental resonance, and frequencies above the fundamental resonance. If we consider a collection of bubbles of a single size, and therefore a single fundamental resonance frequency, the acoustical propagation properties of bubbly water can be summarized from published results of theoretical and experimental studies of bubble screens.

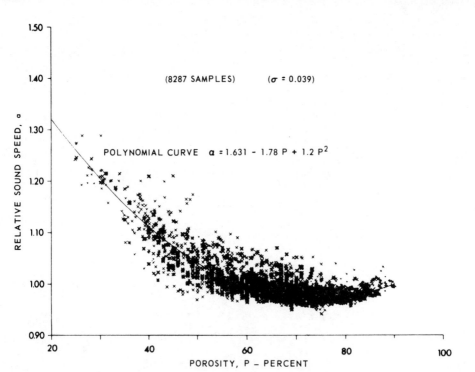

Fig. 4. Relative sound speed (ratio of speed in sediment to speed in water) versus sediment porosity [Akal, 1972].

The attenuation of sound transmitted at normal incidence through such a bubble screen is quite large for frequencies near resonance, but is decreased for off-resonance frequencies. This is illustrated in Figure 5, where theoretical and experimental transmission loss results are shown after Macpherson [1957]. Attenuation measurements are also reported by Cartensen and Foldy [1947] and by the authors given below who report sound speed measurements.

The theoretical speed of propagation of sound through such a screen is illustrated in Figure 6 after Buddruss [1971]. Far below resonance, sound speed is below the bubble-free water sound speed, is independent of frequency, and is accurately described by Wood's emulsion equation [Wood, 1955]. In the vicinity of the resonance frequency, sound speed first decreases and then rapidly increases to a value above that for bubble-free water. For even higher frequencies, the sound speed decreases until it is again frequency independent and is equal to the value for bubble-free water. Confirmation of this theoretical behavior has resulted from measurements

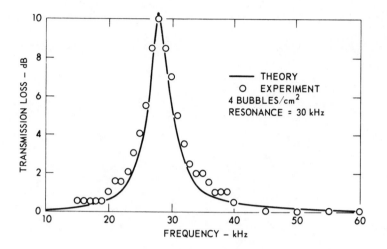

Fig. 5. Normal incidence transmission loss through a screen of bubbles in Water [*Macpherson, 1957*].

reported by *Laird and Kendig* [*1952*]; *Meyer and Skudrzyk* [*1953*]; *Fox et al.* [*1955*]; *Silberman* [*1957*]; *Karplus* [*1958*]; and *Gibson* [*1970*].

The acoustical scattering cross section of an air bubble in water is large near resonance. The work of *Strasberg* [*1956*] indicates that the volume pulsation (lowest order) mode of bubble vibration is the only one which is an efficient radiator. Thus the scattering from a single bubble is omnidirectional. Figure 7 illustrates the frequency dependence of the scattering cross section of a single bubble. The large scattering cross section is one reason for a large attenuation near resonance: energy is scattered out of the propagating wave omnidirectionally (other loss mechanisms include thermal, radiation, and viscous damping of the bubble motion, which are also large near resonance due to the increased bubble wall motion [*Devin, 1959; Eller, 1970*]). Measurements at normal incidence of acoustical reflection from a bubble screen have been reported by *Macpherson* [*1957*]. His results are shown in Figure 8. Maximum reflectivity is observed near resonance. Measurements of side-scattering from bubbly water reported by *Mole et al.* [*1972*] compare favorably with theory and illustrate the difficulty of making this measurement.

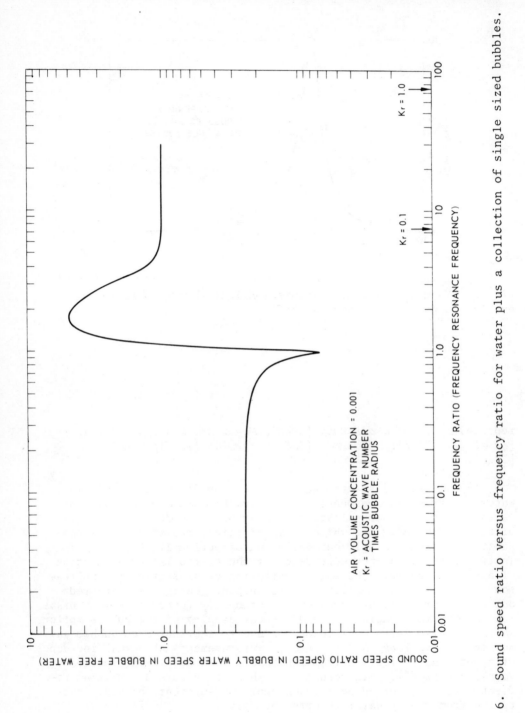

Fig. 6. Sound speed ratio versus frequency ratio for water plus a collection of single sized bubbles.

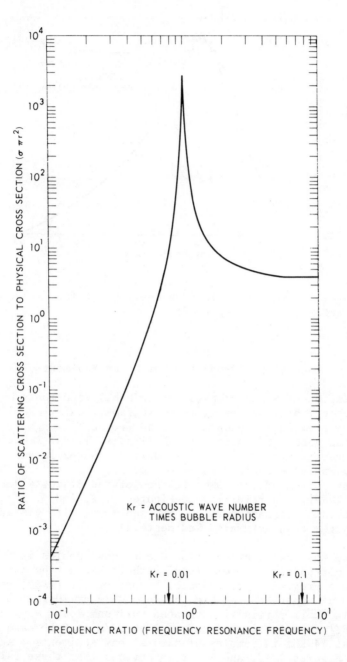

Fig. 7. Normalized scattering cross section versus frequency for a single bubble in water

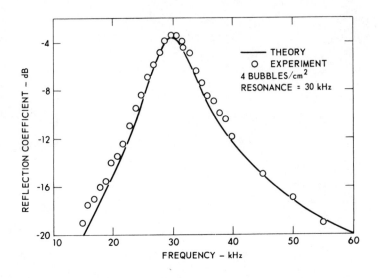

Fig. 8. Normal incidence reflection coefficient for a screen of bubbles in water [*Macpherson, 1957*].

C. Acoustical Properties of Gassy Sediments

Attenuation of sound propagating in bubbly sediments has not been the subject of many measurement programs. *Muir* [*1972*] reports attenuations from 25 to 90 dB/m over the frequency range from 40 to 80 kHz in the gassy mud bottom of Lake Travis, Texas. *Anderson et al.* [*1971*] report attenuations of 13 dB/m at 40 kHz in gassy Lake Austin mud, and attenuations in excess of 130 dB/m at 300 kHz in the gassy bottom of the lower reaches of the Brazos River [*Anderson et al. 1968*]. In referring to Figures 1, 2, and 3, it can be seen that these attenuation values for sediments with gas are only slightly larger than saturated sediment values.

For the most part, attenuation due to gas bubbles in sediments has been reported as an interference rather than as the subject of a measurement program. For example, *Nyborg et al.* [*1950*] report prohibitive attenuations, as high as 2600 dB/m at 10 kHz to 6400 dB/m at 30 kHz in their partially saturated soil samples. *Wood and Weston* [*1964*] report attenuations of 1750 dB/m at 8 kHz and of 2400 dB/m at 14 kHz in samples of natural bay sediments. Again referring to Figures 1, 2, and 3, it is obvious that these values are orders of magnitude greater than for saturated sediments. Both Nyborg et al. and Wood and Weston report significantly decreased values for attenuation after the gas had been drawn from the sediments by placing them in a vacuum chamber (see the data of Wood and Weston for saturated sediments in Figure 1).

ARL was first introduced to the impact of gas bubbles on acoustical propagation during a series of measurements of the acoustical properties of laboratory constructed sediments [*Nolle et al. 1963; Hampton, 1966, 1967*]. Nolle found it necessary to remove air from his laboratory constructed water-sand mixture by boiling in order to prevent measuring anomalously high values for attenuation and interface scattering coefficients.

Hampton, who constructed laboratory sediments for measurement of acoustical properties, used pure kaolinite clay, kaolinite and sand mixtures, and black garden soil. As saturated sediments were desired for these measurements, the samples were placed in a vacuum for a period of several hours to remove entrapped gas. When these precautions were not taken, very large values of attenuation were observed.

One interesting example of naturally generated gas (rather than air entrapped during construction of the sediment) occurred during Hampton's study. This was his first work performed on the problem, and he used a large sample of quasinatural sediment. A 2-meter diameter by 2.3 meters deep steel tank was filled by wet-sieving black garden soil to remove all particles larger than sand size. This large quantity of sediment was suspended at a depth of 5 meters under a floating platform at the ARL Lake Travis Test Station. After about two months, detectable quantities of methane gas were generated in the sediment. The gas was first detected by the fact that the measured attenuation exceeded 260 dB/m (the limit of the measuring technique) for any of the frequencies measured (50 to 200 kHz). The sediment continued to generate generous quantities of gas for over two years, with no measurable reduction in the original 7% of organic content.

Many authors report a greatly decreased sound speed when gas bubbles are present in a sediment [*Jones et al. 1958 and 1964; Brandt, 1960; Levin, 1962; Lewis, 1966;* and *Hochstein, 1970*]. There is, however, some disagreement reported: an increase of sound speed has been observed when bubbles are present (*Brutsaert and Luthin* [*1964*] discuss this). The key to this issue may well be the degree to which bubbles in sediments act as a resonant system, and the relation of the acoustic frequency to the bubble-sediment resonance frequency. If the gassy sediment can be described by a sound speed behavior such as that in Figure 6 for bubbles in water, the controversy could be resolved, and sound speed measurements versus frequency would be revealed as a valuable tool for *in situ* bubble size assessment.

Many studies have shown that gas bearing sediments form highly reflective pressure release boundaries [*Bobber, 1959; Leslie, 1960; Grubnik, 1961; Levin, 1962; Ruff, 1967; Schubel, 1974*]. These measurements are all for frequencies of less than 10 kHz.

The examples given have been for gas bearing sediments in shallow water. Very little published information is available on the acoustical properties of sediments in deep ocean which are known to contain gas. A notable exception is the report of *Stoll et al.* [*1971*], which describes anomalously high sound speeds in a gassy layer of sediment in 3600 meters of water. This high value for sound speed is attributed to the probable occurrence of the gas in the form of a gas hydrate (clathrate).

INVESTIGATION OF GAS IN SEDIMENT

A set of environmental measurements in the lower reaches of the Brazos River, Texas [*Anderson et al. 1968*], conducted in support of acoustical experiments [*Muir et al. 1968a,b*] revealed a highly attenuating, near surface, gassy sediment layer. This finding provoked renewed interest at ARL in the problem of gas-bearing sediments, and resulted in a specific study of some of the features of these sediments [*Anderson et al. 1971*]. The study concentrated on development of a chemical model for a gassy sediment, with concurrent work on sampling techniques and limited acoustical measurements.

Bubbles are one end-product of the system (Figure 9) in which organic matter is incorporated into bottom sediments, decomposed, and released as small organic molecules. A chemical model of this system can be summarized by a set of assumptions and a chemical equation. The assumptions about the bottom sediment system are

1. Bottom sediment is a dynamic chemical system;

2. Materials which enter the sediment are altered, and the products react and become immobilized, or migrate out of the system;

3. Many thermodynamically possible reactions occur, by the way of the catalyzing effect of bacterial metabolism;

4. The bacteria necessary for performing these reactions are always present;

5. Organic-rich bottom sediments are frequently anaerobic;

6. Bottom sediments can be treated chemically as being an approximate equilibrium system;

7. Bubbles will form when interstitial water becomes oversaturated with any gas.

The composite chemical equation which includes the major reactions of the bottom sediment system is

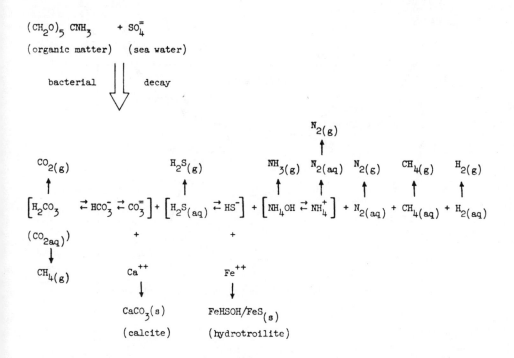

This equation is a qualitative, unbalanced representation of the model system, an expansion of Figure 9 which allows visualization of the results of bacterial decay as specific chemical products. It states that as long as sufficient organic matter is present in the sediment, bacterial decay will continue. If analysis of a sediment shows that organic matter is present, the equation must be operating provided the sediment is anoxic. The predicted product gases N_2, CO_2, H_2, H_2S, and CH_4 have been reported as components of sediment bubbles [Olson and Wilder, 1961; Werner, 1968]. These and others, including NH_3, have been identified in interstitial water samples. The highly soluble NH_3, if present in any bubbles, may have been lost by solution into the water when gas samples were collected.

The relative proportions of the various product species determine the pH and oxidation potential (Eh) of the interstitial water solution. These two parameters can be taken as indicators of the chemical conditions in a sediment. For any dissolved species, a theoretical equilibrium diagram can be constructed, using pH and Eh as variables, which shows the expected predominant species at any combination of Eh-pH conditions. From such a diagram one can

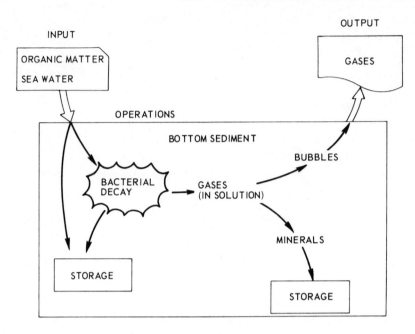

Fig. 9. Bottom sediment biological reaction system

predict which species will be produced in the greatest amount [Garrels and Christ, 1965; Thorstenson, 1969]. As bacterial decay continues, the concentrations of the dissolved species will continue to increase until saturation is reached. The abundance of nucleation sites in sediments results in bubble formation by the first species to reach saturation. Other species in solution will then diffuse into the bubble until each attains diffusion equilibrium across the boundary. As more gas is produced, these bubbles will tend to grow and coalesce until they are able to rise through the sediment and into the overlying water.

The probable gases in bubbles of a given sediment can thus be determined by measuring, or determining from the literature, the Eh and pH of the sediment and by referring to Eh-pH equilibrium diagrams. The number and size of the gas bubbles trapped in a sediment will depend on its grain size, porosity, viscosity, and other mass properties. This dependence can probably be best determined empirically.

The gas sampler, which was developed during the investigation of gassy sediments, is shown diagrammatically in Figure 10. Previous attempts to analyze gas bubbles from sediments have involved the use

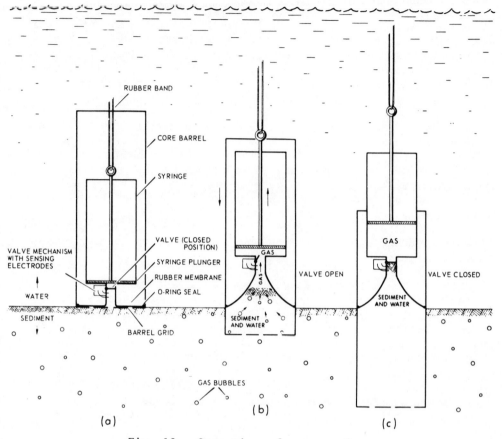

Fig. 10.   Operation of gas sampler.

of an inverted funnel sampling device, with the bubbles rising
through and displacing water in the funnel and collecting container.
This method has the serious disadvantage of allowing the sediment
gases to come into contact with and exchange, to an unknown extent,
with gases in the water they displace. The exchange is probably extensive, as the gases in sediments are quite different from those
that would be found in the overlying water.

Uncontaminated gas bubbles from a known volume of bottom sediment can be collected with the sampling device illustrated in Figure
10. The sampler's principal components are: a core barrel with a
coarse wire grid at the bottom, a rubber membrane which can be formed
into a funnel, a gas-tight 1500 ml syringe, and an electronically
operated gas valve.

Scuba divers operate the sampler by first placing it on the

bottom with the funnel flattened, the valve closed, and the syringe empty (Figure 10a). With the flattened funnel flush with the sediment-water interface, the only water between the collecting device and the bubbles to be collected is interstitial sediment water. The gases in solution in this water are assumed to be in equilibrium with the bubbles. Thus, only small exchange can occur, and the bubble composition will be unchanged during sampling.

In the next step, divers force the core barrel down into the sediment by pushing on handles attached to the barrel. As they begin coring, the funnel is formed (Figure 10b). Bubbles in the sediment inside the core barrel are jarred loose by the core barrel grid. They rise through the loosened sediment to collect in the funnel. The valve opens when the gas reaches the funnel apex and breaks electrical contact (water connection) between two implanted electrodes. The gas then passes into the syringe, aided by a slight suction produced by a rubber band pulling upward on the syringe plunger (Figure 10b).

Use of the automatic electric valve has several advantages. It prevents water from entering the syringe by closing whenever water makes contact between the sensing electrodes (response time is about 10 msec). Because of its rapid response, it will collect even small bubbles as they rise into the funnel. Also, it operates without the diver's needing to see or handle it. The use of the syringe as a collecting reservoir has two fundamental advantages. First, the gas is displacing a solid plunger rather than a liquid with which it could exchange gases. Second, once a sample is taken, gas is not lost by expansion as the sampler is raised to the surface.

MEASUREMENT OF SOUND SPEED WHILE CORING

Sediment acoustical and mechanical properties, which are measured in the laboratory for either constructed sediments or cores, require temperature and pressure correction to determine their *in situ* values. Limits to the capability of obtaining undisturbed samples for such measurements contribute to uncertainty about actual *in situ* values. *Hamilton* [*1971*] has given a technique for extrapolating laboratory sound speed measurements to *in situ* values. His method is based on laboratory and *in situ* measurements of surface sediments of the North Pacific. Extending this technique to other regions or to sediments deeper than the surface layer remains to be tested. Furthermore, the presence of gas either as bubbles or as a hydrate (clathrate) complicates any attempt to extrapolate to sea floor conditions. *In situ* measurements of sound speed by probing have previously been limited to surface sediments [*Hamilton, 1963; Bennin and Clay, 1967; Lewis et al. 1970*].

Sediment cores are taken with core lengths of 20 meters or more

in the bottom. ARL has developed a profilometer system for attachment to corers in order to obtain an *in situ* sediment sound speed
profile during coring operations. The system uses two piezoelectric
transducers mounted in the cutting head of the corer and is connected
to the electronics housing by coaxial cables. An acoustical pulse
is generated by the transmit transducer. This pulse traverses the
sediment contained in the cutting head at that instant and is received by the second transducer that is diametrically across the
cutting head from the projector. The time delay resulting from the
pulse traversing the sediment in the cutter head is converted to a
voltage output that is presently sent to the surface by a cable and
recorded on magnetic tape. This output voltage is calibrated to
give a precise indication of the speed of propagation of sound in
the sediment at the pulse carrier frequency. When used with a free-
falling corer (either a gravity or piston corer), an accelerometer
is mounted in the watertight electronics box; its output is also
available for recording to aid in interpreting the profile information.

Figure 11 presents an example of a sound speed profile taken
with this system in a homogeneous sediment of the Mississippi Delta
region of the Gulf of Mexico. This record was obtained with the
profilometer attached to the piston corer of Texas A&M University.
Using a published summary [*Akal, 1972*] that relates sound speed and
porosity, together with measured porosities in the core, the sediment sound speed for the core of Figure 3 should be about 15 m/sec
less than the bottom water sound speed and should vary less than
±15 m/sec along the core; these figures are in close agreement with
the measurements.

Another *in situ* sound speed profile and associated description
of core lithology is shown in Figure 12 for a portion of a core obtained in Baffin Bay, Texas. Gas bubbles of about 1 mm diameter were
sparsely distributed throughout the upper few centimeters of the core.
The measured sound speed in the sediment was less than that in water.
The decreased speed was, however, within the bounds [*Akal, 1972*] to
be expected for this high porosity sediment if it were saturated with
water. Thus, at the 190 kHz carrier frequency of the profilometer,
the sound speed in this sediment is not significantly changed by the
size and quantity of bubbles it contains.

Attempts to use the profilometer to obtain sound speed profiles
in the sediments of Lake Travis, Texas, have been unsuccessful because the sediment attenuation was greater than the profilometer
automatic gain control could compensate. Cores of the sediment revealed a large gas content, so much that the sediment samples had
a spongy appearance. Another nearby region, Lake Bastrop, was probed
with short acoustical probes to determine the feasibility of testing
the profilometer there. Water sound speed was 1470 m/sec. Sound
speed and attenuation at 110 kHz for three depths in the sediment

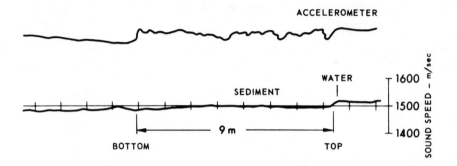

Fig. 11.   Sound speed profile, Mississippi delta, Gulf of Mexico

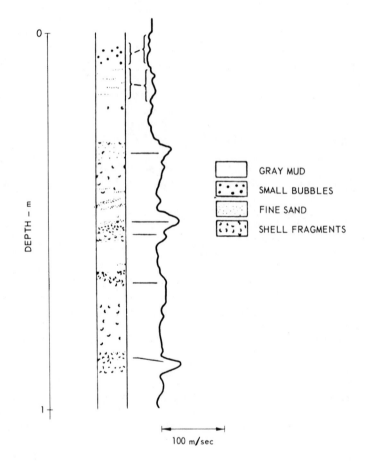

Fig. 12.   Comparison of sound speed profile and lithology of the upper portion of a core from Baffin Bay, Texas.

were measured to be about: 7.5 cm depth, 1250 m/sec, 105 dB/m;
15 cm depth, 1270 m/sec, 420 dB/m; and 22.5 cm depth, 1220 m/sec,
475 dB/m. Although neither physical properties nor gas content were
measured for this sediment, it was observed that gas bubbles were
released when the sediment was agitated. The fact that attenuation
increased directly with depth, while sound speed first increased
and then decreased, supports the assumption that there are complex
relationships between bubble size and concentration and acoustical
parameters (sound speed, attenuation, and frequency).

EXPANSION OF PROFILOMETER CAPABILITIES

There is an ongoing study of the feasibility of expanding the
profilometer capabilities to include a measure of sediment acoustic
attenuation and internal volume scattering.

The automatic gain control feedback voltage is a measure of
the insertion loss of the material between the profilometer transducers at any instant. This voltage has been separately recorded
in a limited number of coring tests. Figure 13 compares profiles
of sound speed and relative received signal amplitude (as measured
by the AGC feedback voltage) for a single core from Baffin Bay,
Texas. Increased sound speed in these cores is well correlated with
increased sand content. As Figure 13 demonstrates, the increased
sound speed layers measured are also well correlated with decrease
in the received signal amplitude. Thus, though work has not yet
been devoted to reducing this information to a numerical measure of
acoustical attenuation, the expected increase of attenuation with
sand content is observed.

The internal acoustical volume scattering strength of sediments
is not documented in the literature. To test the feasibility of
adding measurement of this parameter to the profilometer, a third
transducer was added to the cutter head. This transducer was placed
so that its acoustical axis intersected at $90°$, the primary acoustical path joining the sound speed measurement transducer pair.
Using the third transducer as a receiver, acoustical energy, which
is side scattered out of the principal transmission path in the
sediment, would be measured.

The acoustical feedover to the volume scatter transducer in
clear water was measured to be about 14 dB below the direct path
signal. Insertion of the cutter head into saturated sand sediment
and muddy sand sediments in the laboratory indicated that the volume
scattered levels were below the acoustical feedover levels. This
measurement will therefore require greater isolation between the
two acoustical paths.

An encouraging indication was obtained from a series of

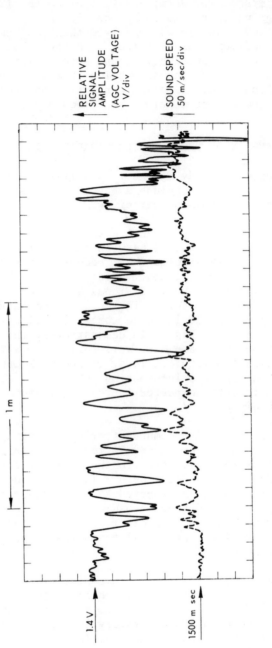

Fig. 13. Relative signal amplitude and sound speed profiles for a core from Baffin Bay, Texas

measurements with the 3-transducer arrangement in water. Gas bubbles, generated by hydrolysis, were allowed to rise into the region which is common to both acoustical paths. Side scattered acoustical levels were above the acoustical feedover for some bubble sizes and concentrations. Thus, a sufficient reduction of the feedover level by change of frequency, different transducer arrangement, or acoustical baffling should allow measurement of volume scattering in sediment. Also, the high levels of scattering from bubbles in water suggest the possibility of internal volume scattering measurements serving as useful indications of gas bubbles in the sediment.

ACKNOWLEDGMENTS

The sampler and chemical model of a gassy sediment are primarily the results of work by R.J. Harwood. The profilometer was designed and built by D.J. Shirley. The work described was supported in part by the Office of Naval Research, Contract N00014-73-C-0362, and by the Naval Ship Systems Command.

REFERENCES

Akal, T., The relationship between the physical properties of underwater sediments that affect bottom reflection, *Mar. Geol.*, *13*, 251, 1972

Anderson, A. L., T. G. Muir, Jr., R. S. Adair, and W. H. Tolbert, A geoacoustic survey of the Brazos River, Part I: Environmental studies, *Def. Res. Lab. Acoust. Rep. 294* (DRL-A-294), Appl. Res. Lab., The University of Texas at Austin, 1968.

Anderson, A. L., R. J. Harwood, and R. T. Lovelace, Investigation of gas in bottom sediments, *Tech. Rep. 70-28* (ARL-TR-70-28), Appl. Res. Lab., The University of Texas at Austin, 1971.

Bennett, L. C., Jr., *In situ* measurements of acoustic absorption in unconsolidated sediments (Abstract), *Trans. Am. Geophys. Union*, *48*, 144, 1967.

Bennin, R. S., and C. S. Clay, Development of an *in situ* sediment velocimeter, *Tech. Rep. 131*, Hudson Lab., Columbia University, New York, 1967.

Bobber, R. J., Acoustic characteristics of a Florida lake bottom, *J. Acoust. Soc. Amer.*, *31*, 250, 1959.

Brandt, H., Factors affecting compressional wave velocity in unconsolidated marine sand sediments, *J. Acoust. Soc. Amer.*, *32*, 171, 1960.

Briggs, H. B., J. B. Johnson, and W. P. Mason, Properties of liquids at high sound pressure, *J. Acoust. Soc. Amer.*, *19*, 664, 1947.

Brutsaert, W., and J. N. Luthin, The velocity of sound in soils near the surface as a function of the moisture content, *J. Geophys. Res.*, *69*, 643, 1964.

Buddruss, C. P., Experimentelle Untersuchungen zur Koharenz von Durchgangs und Ruckstreuschall einer Luftblasen-Wasser-Schicht, *Acustica*, *24*, 147, 1971.

Busby, J., and E. G. Richardson, The absorption of sound in sediments, *Geophysics*, *22*, 821, 1957.

Cartensen, E. L., and L. L. Foldy, Propagation of sound through a liquid containing bubbles, *J. Acoust. Soc. Amer.*, *19*, 481, 1947.

Devin, C., Jr., Survey of thermal, radiation, and viscous damping of pulsating air bubbles in water, *J. Acoust. Soc. Amer.*, *31*, 1654, 1959.

Eller, A. I., Damping constant of pulsating bubbles, *J. Acoust. Soc. Amer.*, *47*, 1469, 1970.

Faas, R. W., Analysis of the relationship between acoustic reflectivity and sediment porosity, *Geophysics*, *34*, 546, 1969.

Fox, F. E., S. R. Curley, and G. S. Larson, Phase velocity and absorption measurements in water containing air bubbles, *J. Acoust. Soc. Amer.*, *27*, 534, 1955.

Garrels, R. M., and C. L. Christ, *Solutions, Minerals, and Equilibria*, p. 450, Harper and Row Publishers, New York, 1965.

Gibson, F. W., Measurement of the effect of air bubbles on the speed of sound in water, *J. Acoust. Soc. Amer.*, *48*, 1195, 1970.

Grubnik, N. A., Investigation of the acoustic properties of underwater soil at high acoustic frequencies, *Sov. Phys. Acoust.*, *6*, 447, 1961.

Hamilton, E. L., Sediment sound velocity measured *in situ* from bathyscaph TRIESTE, *J. Geophys. Res.*, *68*, 5991, 1963.

Hamilton, E. L., Reflection coefficients and bottom losses at normal incidence computed from Pacific sediment properties, *Geophysics*, *35*, 995, 1970.

Hamilton, E. L., Prediction of *in situ* acoustic and elastic properties of marine sediments, *Geophysics*, *36*, 266, 1971.

Hamilton, E. L., Sound attenuation in marine sediments, *Ocean Sciences Dept.*, *Rep. NUC TP 281*, Naval Undersea Center, San Diego, California, 1972.

Hampton, L. D., Acoustic properties of sediments, *Defense Res. Lab. Acoust. Rep. 254* (DRL-A-254), Defense Res. Lab., The University of Texas, 1966.

Hampton, L. D., Acoustic properties of sediments, *J. Acoust. Soc. Amer.*, *42*, 882, 1967.

Hochstein, M. P., Seismic measurements in Suva Harbour (Fiji), *N. Z. J. Geol. Geophys.*, *13*, 269, 1970.

Houghton, G., Theory of bubble pulsation and cavitation, *J. Acoust. Soc. Amer.*, *35*, 1387, 1963.

Jones, J. L., C. B. Leslie, and L. E. Barton, Acoustic characteristics of a lake bottom, *J. Acoust. Soc. Amer.*, *30*, 142, 1958.

Jones, J. L., C. B. Leslie, and L. E. Barton, Acoustic characteristics of underwater bottoms, *J. Acoust. Soc. Amer.*, *36*, 154, 1964.

Karplus, H. B., The velocity of sound in a liquid containing gas bubbles, *Armour Res. Found. Rep.* COO-248, Illinois Inst. of Tech., 1958.

Laird, D. T., and P. M. Kendig, Attenuation of sound in water containing air bubbles, *J. Acoust. Soc. Amer.*, *24*, 29, 1952.

Leslie, C. B., Normal incidence measurement of acoustic bottom constants, *U. S. Naval Ord. Lab. NAVORD Rep. 6832*, White Oak, Silver Spring, Maryland, 1960.

Levin, F. K., The seismic properties of Lake Maracaibo, *Geophysics*, *27*, 35, 1962.

Lewis, L. F., Speed of sound in unconsolidated sediments of Boston Harbor, Massachusetts, M. S. thesis, Mass. Inst. of Tech., Cambridge, Massachusetts, 1966.

Lewis, L. F., V. A. Nacci, and J. J. Gallagher, *In situ* marine sediment probe and coring assembly, *U. S. Naval Underwater Sound Lab., NUSL Rep. 1094*, New London, Connecticut, 1970.

Lewis, L. F., An investigation of ocean sediments using the deep ocean sediment probe, Ph.D. thesis, Univ. of Rhode Island, Kingston, 1971.

Macpherson, J. D., Effect of gas bubbles on sound propagation in water, *Proc. Phys. Soc. (London)*, *B 70*, 85, 1957.

McCann, C., and D. M. McCann, The attenuation of compressional waves in marine sediments, *Geophysics*, *34*, 882, 1969.

McLeroy, E. G., and A. DeLoach, Sound speed and attenuation, from 15 to 1500 kHz, measured in natural sea-floor sediments, *J. Acoust. Soc. Amer.*, *44*, 1148, 1968.

Meyer, E., and E. Skudrzyk, Uber die akustischen Eigenschaften von Gasblasenschleiern in Wasser, *Acustica*, *3*, 434, 1953.

Minnaert, M., On musical air bubbles and the sounds of running water, *Phil. Mag.*, *10*, 235, 1933.

Mole, L. A., J. L. Hunter, and J. M. Davenport, Scattering of sound by air bubbles in water, *J. Acoust. Soc. Amer.*, *52*, 837, 1972.

Muir, T. G., Jr., J. G. Pruitt, R. S. Adair, and J. E. Blue, A geoacoustic survey of the Brazos River, Part II: Ultrasonic attenuation studies, *Defense Res. Lab. Acoust. Rep. 294* (DRL-A-294), Appl. Res. Lab., The University of Texas at Austin, 1968a.

Muir, T. G., Jr., R. S. Adair, J. G. Pruitt, and J. G. Willette, A geoacoustic survey of the Brazos River, Part III: Reverberation studies, *Defense Res. Lab. Acoust. Rep. 294* (DRL-A-294), Appl. Res. Lab., The University of Texas at Austin, 1968b.

Muir, T. G., Experimental capabilities of the ARL sediment tank facility in the study of buried object detection, *Appl. Res. Lab. Tech. Memo. 72-32* (ARL-TM-72-32), The University of Texas at Austin, 1972.

Nolle, A. W., W. A. Hoyer, J. F. Mifsud, W. R. Runyan, and M. B. Ward, Acoustical properties of water-filled sand, *J. Acoust. Soc. Amer.*, *35*, 1394, 1963.

Nyborg, W. L., I. Rudnick, and H. K. Schilling, Experiments on acoustic absorption in sand and soil, *J. Acoust. Soc. Amer.*, *22*, 422, 1950.

Olson, F. C. W., and B. Wilder, Gases in bottom sediments, *Bull. Marine Sci. Gulf and Caribbean*, *11*, 207, 1961.

Ruff, G. A., Acoustic characteristics of Black Moshannon Lake bottom, *J. Acoust. Soc. Amer.*, *42*, 524, 1967.

Schirmer, F., Schallaus breitung im Schlick, *Deut. Hydrogr. Z.*, *23* (1), 24, 1970.

Schubel, J. R., Gas bubbles and the acoustically impenetrable, or turbid, character of some estuarine sediments, in *Natural Gases in Marine Sediments*, edited by I. R. Kaplan, pp. 275-298, Plenum Press, New York, 1974.

Shima, A., The natural frequency of a bubble oscillating in a viscous compressible liquid, *J. Basic Eng.* (Trans. ASME), 555, 1970.

Shumway, G., Sound speed and absorption studies of marine sediments by a resonance method, Part I, *Geophysics*, *25*, 451, 1960.

Silberman, E., Sound velocity and attenuation in bubbly mixtures measured in standing wave tubes, *J. Acoust. Soc. Amer.*, *29*, 925, 1957.

Stoll, R. D., J. Ewing, and G. M. Bryan, Anomalous wave velocities in sediments containing gas hydrates, *J. Geophys. Res.*, *76*, 2090, 1971.

Strasberg, M., Gas bubbles as sources of sound in liquids, *J. Acoust. Soc. Amer.*, *28*, 20, 1956.

Thorstenson, D. C., Equilibrium distribution of small organic molecules in natural waters, Ph.D. thesis, Northwestern University, Chicago, 1969.

Ulonska, A., Versuche zur Messung der Schallgeschwindig keit und Schalldampfung im Sediment *in situ*, *Deut. Hydrogr. Z.*, *21* (2), 49, 1968.

Urick, R. J., The absorption of sound in suspensions of irregular Particles, *J. Acoust. Soc. Amer.*, *20*, 283, 1948.

Werner, A. E., Gases from sediments in polluted coastal waters, *Pulp and Paper Magazine of Canada*, *69*, 127, 1968.

Wood, A. B., *A Textbook of Sound*, pp. 360-363, New York, 1955.

Wood, A. B., and D. E. Weston, The propagation of sound in mud, *Acustica*, *14*, 156, 1964.

GAS BUBBLES AND THE ACOUSTICALLY IMPENETRABLE, OR TURBID, CHARACTER OF SOME ESTUARINE SEDIMENTS

J.R. Schubel

Chesapeake Bay Institute
The Johns Hopkins University
Baltimore, Maryland 21218

ABSTRACT

Many shallow water, fine-grained sediments apparently contain appreciable quantities of interstitial gas bubbles produced by the biochemical degradation of organic matter. There have been relatively few direct observations of entrapped bubbles, but the presence of extensive gaseous sedimentary zones has frequently been inferred from high resolution sub-bottom profiling records. Frequently, large segments of many records made in estuaries and bays with high-frequency, low-energy profilers are confused and characterized by bands of strong diffuse reflection that mask the underlying features. Investigators have repeatedly attributed this anomalous acoustic behavior to the presence of interstitial gas bubbles, which produce excessive reverberation of sound within the sediment. Until recently, however, this hypothesis for the explanation of the acoustically impenetrable or 'turbid' character of sediments had apparently not been tested.

Acoustically 'turbid sediments' cover large areas of Chesapeake Bay. Along a cross-section parallel to the Bay Bridge near Annapolis, acoustically turbid zones alternate with zones where good sub-bottom records can be obtained. The sediments are quite similar in grain size, mineralogy, and water content. Determinations of the compressibility of sediment samples from the acoustically turbid zones, and from contiguous zones where good sub-bottom records were obtained, showed that the acoustically turbid sediments are several orders of magnitude more compressible than the acoustically clear sediments. The increased compressibility is a result of the presence of interstitial gas bubbles. X-ray examination of frozen cores confirmed this. Bubbles were scarce or absent in cores from

acoustically clear areas. Although the acoustically turbid zones
are of considerable horizontal extent, their vertical dimensions
may be only 1 to 3 meters.

The cause of the acoustically turbid character of some other
zones in Chesapeake Bay is buried shell beds (bars). The compressibility of sediments from these areas is not anomalously high.
The two kinds of acoustically turbid zones can normally be distinguished by a careful examination of the sub-bottom profiling records.

High resolution sub-bottom profilers can be used to map the
distribution of gas bubble zones, and for the selection of coring
stations for gas analysis.

INTRODUCTION

Many shallow water, fine-grained sediments, particularly in
bays and estuaries, are almost acoustically impenetrable to the
energy from high resolution continuous seismic profilers with peak
outputs of up to 2-3 kJ at frequencies of 0.2-10 kHz. The records
from such areas are confused and characterized by bands of strong
diffuse reflection which mask the underlying features (Figures 1
and 2). *Schubel and Schiemer* [*1972*] proposed calling such zones
'acoustically turbid' sediments, a name that is descriptive of their
appearance on sub-bottom profiling records, and a name that does
not have a genetic connotation. Acoustically 'turbid' sediments
have been reported in large areas of the Chesapeake Bay and its
tributaries [*Stiles, 1970; Grim et al. 1970; Schubel and Zabawa,
1972*], in Delaware Bay [*Moody and van Reenan, 1967;* R. Moose, personal communication, 1971], in the Hudson River (Alan Bieber, personal communication, 1970), in Long Island Sound [*Grim et al. 1970*],
in Australian estuaries, in channels of the Mekong Delta in South
Vietnam [*Stiles et al. 1969*], in a coastal pond in Massachusetts
[*Emery, 1969*], in the Sea of Galilee in Israel [*Klein and Edgerton,
1968*], and in many other lakes, rivers, bays and estuaries. Since
the majority of shallow water profiling has been done on a con-
sulting basis by private service corporations, most of the results
are unpublished and the extent of acoustically turbid sediments may
be largely unknown. They are, however, widespread in their distribution, and their presence makes the determination of the shallow sub-
structure both difficult and expensive since borings must be made.
Knowledge of the distribution of acoustically turbid sediments and
of the mechanisms responsible for their formation may be important
militarily.

Investigators have repeatedly referred to acoustically turbid
sediments as 'organic' or 'gaseous zones', and have ascribed their
anomalous acoustic behavior to the presence of interstitial gas

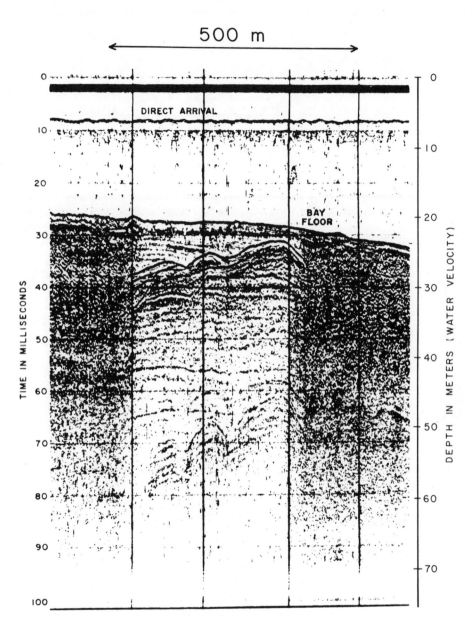

Fig. 1. Photograph of a high resolution continuous seismic profiling record made with an E.G.&G. High Resolution Boomer showing a zone of good penetration flanked by acoustically turbid zones. The record was made near the Chesapeake Bay Bridge (Annapolis, Md.). The zone of good penetration is just downstream from Bridge pier 31.

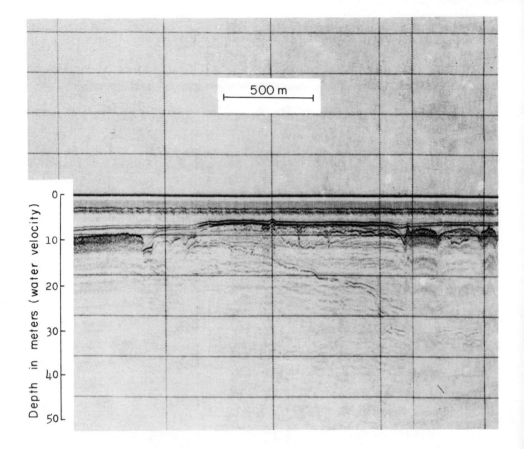

Fig. 2. Photograph of a high resolution continuous seismic profiling record showing both acoustically turbid zones and zones of
good penetration.

bubbles. Although generally not stated, it is implied that the
bubbles are CH_4, CO_2, and H_2S produced by the biochemical degradation of organic matter. The formation and entrapment of gas
bubbles in organic-rich, fine-grained sediment is not unexpected
[Reeburgh, 1969; Anderson et al. 1968], and it is well known that
the acoustical properties of sediment can be drastically altered
by the presence of interstitial gas bubbles as bubbles, even in relatively small quantities, can markedly reduce the speed of sound
and increase its attenuation and reverberation [Nyborg et al. 1950;
Jones et al. 1958, 1964; Bobber, 1959; Levin, 1962; Wood and Weston, 1964; Hampton, 1966; Anderson et al. 1968].

Observations of the effects of gas bubbles on the acoustical properties of sediments have been made both in the laboratory and in the field. Acoustical measurements in areas where the ebullience of bubbles from the bottom has been observed indicate that the surficial sediments are characterized by high reflectivities and low sound velocities, and that the bottom acts as a pressure release surface [Jones et al. 1958, 1964; Bobber, 1959; Levin, 1962]. Until recently, however, apparently no attempt had been made to compare the gas bubble contents of sediment samples from acoustically turbid areas with the bubble contents of samples from contiguous areas where good sub-bottom records were obtained.

In 1970, the Chesapeake Bay Institute initiated a program to study the acoustically turbid sediments near the Chesapeake Bay Bridge. The objectives of the study were to determine whether the acoustically turbid sediments have greater volumes of entrapped gas bubbles than do the adjacent areas where good sub-bottom records were obtained, to determine if there are other important physical differences between acoustically turbid and acoustically clear sediments, and to determine something of the vertical extent of the acoustically turbid sedimentary layers.

This paper is a summary of the findings of that study. Some of the data were presented earlier in the 'grey' literature [Schubel and Schiemer, 1972], and will be published within a few months [Schubel and Schiemer, in press]. These data are reviewed in this report, and some new data are presented.

STUDY AREA

The problem of acoustically turbid sediments is clearly revealed in the fine-grained sediments of large areas of Chesapeake Bay. Continuous seismic profiles in the vicinity of the Chesapeake Bay Bridge, which connects the western shore of Maryland near Annapolis with Kent Island on the Eastern Shore (Figure 3), reveal alternating zones of good penetration and acoustically turbid zones in which energy is strongly reflected in the upper few meters completely masking the underlying features (Figure 4). A number of displacement type devices (boomers) and low energy (up to 3 kJ) sparkers have been used in this area over the past few years, and all have failed to yield penetrations of more than a few meters in the acoustically turbid sediments.

The measurements for the study were made over nearly the entire length of the cross-section near the Bridge, but the most intensive work was done in the areas near piers 22, 28 (acoustically turbid zones), and 31 (an acoustically clear zone). The water depth is 12 meters at pier 22, 17 meters at pier 28, and 18 meters at pier 31.

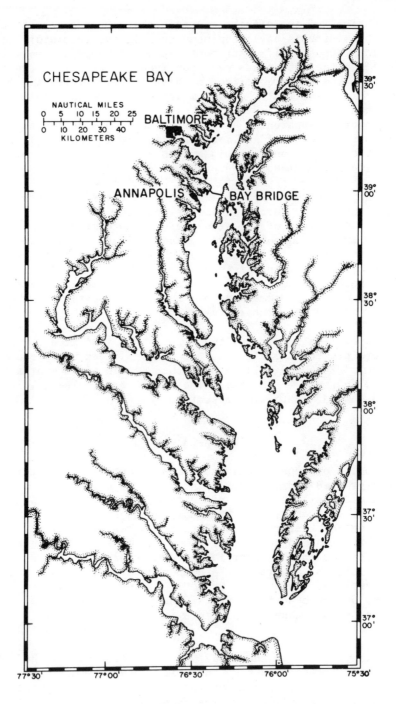

Fig. 3. Map of Chesapeake Bay showing the study area near the Bay Bridge.

Fig. 4. Line drawing of continuous seismic profiling records near the Chesapeake Bay Bridge showing the alternating zones of good penetration and acoustically 'impenetrable' sediments. The Bridge pier numbers are shown at the top of the figure.

RESULTS AND DISCUSSION

At the beginning of the study, an attempt was made to directly compare the volumes of gas bubbles entrapped in the interstices of acoustically clear and acoustically turbid sediments from similar water depths. A special gas tight 'bubble valve' was constructed that attaches to the top of a core liner, and is used in a gravity corer. The device was designed to capture the gas bubbles, which, because of the decrease in pressure, would presumably be released during recovery of the core. The volume of gas could then be measured on deck and related to core length, and its composition could be determined by gas chromatography. The water depths at the stations were similar, 12 meters at pier 22, 17 meters at pier 28, and 18 meters at pier 31, but corrections for differences in the *in situ* pressure could easily be made.

The initial results were encouraging. At the first station, pier 28, an acoustically turbid zone, a large volume of gas (> 50 ml) was released from the sediment before the corer was brought on deck, and the ebullience continued for some time. The gas was primarily methane. At the second station, pier 31, an acoustically clear zone, no gas bubbles were released from the sediment even after the core had been on deck for more than 30 minutes. But, at the third station, pier 22, a very acoustically turbid zone, no gas bubbles were released from a 1.7 meters core, even after vigorous beating of the core in its liner with a piece of rubber hose, and after more than one-half hour of intense agitation with a modified chipping hammer. The project had been complicated immensely by making 'one too many stations'. If interstitial gas bubbles are the cause of the acoustically turbid character of the sediments, both at pier 22 and at pier 28, an explanation of the apparent difference in their gas bubble contents is required.

This apparent discrepancy may be a result of the difference in the permeability of the sediment from the two areas. The sediment near pier 28 is much coarser (silty sand) than the sediment at pier 22 (silty clay), and is much more permeable. Gas bubbles are obviously able to migrate freely through the sediment from pier 28. At pier 22, however, if interstitial gas bubbles are present, they are apparently not able to migrate through the sediment because of its lower permeability. The release of gas bubbles was not observed in cores taken from two other fine-grained areas near piers 20 and 36, both of which are very acoustically turbid areas. A different approach was required to determine if gas bubbles are present in the fine-grained, acoustically turbid zones.

Short ($\leq$ 2 meters) gravity cores were taken at piers 22 and 31 with a Benthos corer. Some of the cores were frozen in liquid nitrogen immediately after being brought on deck; the remainder were refrigerated on the ship. The frozen cores were sliced and X-rayed while still frozen for examination for gas bubbles. The other cores were kept refrigerated until analyzed. It was previously reported by *Schubel and Schiemer* [*1972*] that, "There was extensive evidence of worm burrowing at pier 22, evidence which was lacking at pier 31, and some evidence of the presence of a few spherical cavities (gas bubbles?) at pier 22, but the evidence was not convincing". Further examination of the original X-ray radiographs and of more recent data indicates that spherical cavities, almost certainly gas bubbles, are more abundant at pier 22 than originally reported (Figure 5). It also appears that a more likely explanation of the origin of the small fissures (see Figure 5), which are so abundant in the pier 22 cores, is that they are produced by expansion of gas bubbles during recovery of the corer, and not by burrowing as originally suggested by *Schubel and Schiemer* [*1972*]. The preferential vertical orientation of these features is probably a result of the sampling; the core liner confines the sample laterally, but the upper and lower sediment surfaces are not constrained. There are relatively few spherical cavities and fissures in the upper 25 cm of the cores from pier 22, but they increase rapidly with depth, at least to about 1.75 meters, which is the maximum length of the cores. This distribution pattern agrees well with the compressibility measurements described later (see Figure 9).

To test this proposed hypothesis for the origin of these fissures, a core should be taken at pier 22, placed in a pressurized container *in situ* and maintained at the *in situ* pressure during recovery and X-raying. No evidence of bubbles was found in any of the cores from pier 31, the acoustically clear zone (Figure 6).

The X-ray radiographs provide convincing qualitative evidence of the difference in the gas bubble contents of the sediment in an acoustically turbid area (pier 22) and the sediment in an acoustically clear area (pier 31) where good sub-bottom records were obtained. In an attempt to get a more quantitative indicator of the difference in the gas bubble contents of these two 'kinds' of sediments, another indicator of the presence of gas bubbles was selected, that of the compressibility of the sediment. The purpose of the tests was not to attempt to determine, with a high degree of accuracy, the compressibility or the bulk modulus of the sediment samples, but rather, to obtain a relatively precise indicator of the compressibility to serve as a comparative index of the gas bubble content.

As we are dealing with a three-phase system (sediment, water, gas), the compressibility that is measured is a composite measure of the compressibility of the liquid, gas, and solid phases. The relative importance of each of the phases depends upon its individual compressibility and the volume of each that is present in the sample. The compressibility of water is about 50×10^{-6} atm^{-1}, while compressibilities as low as 2 to 2.5×10^{-6} atm^{-1} have been reported for fine-grained sediments (clay and clayey silt) [*Skempton, 1961*].

5 cm

Fig. 5. Prints of X-radiographs of sections of a core from pier 22. Note the spherical cavities along the sides of the core, and the numerous fissures.

An increase of the compressibility of an actual sediment sample over that of water is an indication of the presence of gas bubbles in the sediment.

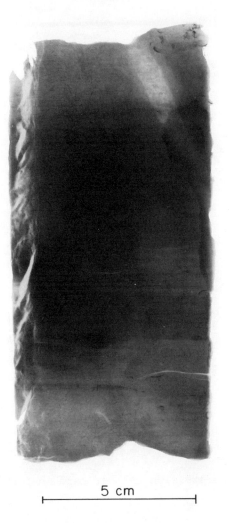

5 cm

Fig. 6. Print of an X-radiograph of a core from pier 31. Note the absence of spherical cavities and fissures.

A compression chamber was constructed from a 36 cm (14 inches) length of 10 cm (4 inches) schedule 80 aluminum pipe (Figure 7). The pipe was capped at the bottom, and fitted with an 'O' ring seal at the top. The cover carried a calibrated gauge glass, and was attached to the pipe with two bolts. The dome of the cap was inclined to prevent accumulation of bubbles during filling. The valve at the bottom was used for filling the testing chamber with water, and the valve at the top served as a gas vent. A connection was also provided at the top for attachment to a nitrogen bottle for pressurization.

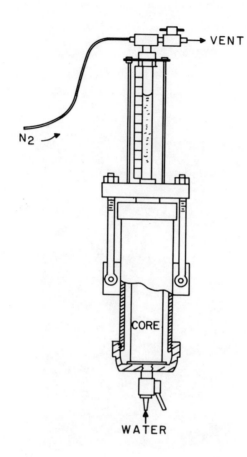

Fig. 7. Sketch of compression chamber used in testing sediment samples.

     To determine the compressibility of a sample, a section of core and liner was cut from a core immediately after it was recovered and carefully inserted into the partially filled testing chamber. The cover was placed on the chamber, and tightened sufficiently to ensure that there would be metal to metal contact at the seal surfaces when the chamber was pressurized. Water was slowly admitted to fill the chamber and the gauge glass to the zero mark. The lower valve was then closed and the pressure in the chamber increased.

     The depression of the water level in the gauge glass with pressurization is a composite measure of the compressibility of the core sample and the surrounding water, and of the change in volume of the testing chamber. The precision of the measurements of the height of the water column is about 1 mm. This corresponds to a volume of about 0.135 ml and to an apparent compressibility of approximately 30-120 x 10^{-6} atm^{-1}, where the value of 30 x 10^{-6} atm^{-1} corresponds to a change of 1 mm in the water level at an increase in pressure of 4 atm, and 120 x 10^{-6} atm^{-1} relates to a change of 1 mm at an increase of 1 atm. The results of tests run with water alone were used to calculate the changes in the volume of the chamber with pressure. The actual sample readings were corrected for the compressibility of the filling water, and for the changes in the volume of the testing chamber. For each sediment sample, the water level was recorded for every 10 psi ($\simeq$ 0.7 atm) increase in pressure to 60 psi ($\simeq$ 4 atm). Readings were also made with every 10 psi decrease in pressure to determine if the volume of the core segment returned to the volume it had at the same pressure before having been subjected to higher pressures. The readings were in close agreement in all but one case. For one sample from pier 22, gas bubbles were evolved from the sediment when the pressure was decreased; the sample, of course, did not return to its original volume.

Cores were taken at piers 22, 25, 28, and 31 with a gravity corer that uses plastic liners with an I.D. of about 6.68 cm (2.63 inches) and an O.D. of about 7.14 cm (2.81 inches). The lengths of the cores ranged from 135 cm to 179 cm. Immediately after a core was taken, a segment of the core and liner was cut off and inserted into the pressure chamber for testing. This was repeated until the entire core had been analyzed. At each station, two to five cores were tested. The agreement between measurements on replicate cores was within the precision of the measurements.

The measurements for a sample from a depth of 120-150 cm in a core from pier 22 are depicted in Figure 8. The results are expressed as the change in the volume of the core segment (ΔV) at different pressures relative to the initial volume of the core segment V_0. The same information for a water core is given for reference. Figure 8 clearly shows that the sediment sample from pier 22, an acoustically turbid zone, was approximately 2 orders of magnitude more compressible than the water core. The measurements of a sample from pier 31, a zone of good penetration, were markedly different from those of the pier 22 sample. The corrected values of $\Delta V/V_0$ for a sample from 100-120 cm depth in a pier 31 core were less than the sensitivity of the apparatus at all tested pressures. The *in situ* pressure at the sediment sample depth was approximately 1.4 atm at pier 22 and approximately 2.0 atm at pier 31.

The $\Delta V/V_0$ at 2 atm of the sediment in the upper 1.5 meters at piers 22 and 31 are shown in Figure 9. The compressibility of the sediment at pier 22 was high throughout the core, and increased monotonically with depth. At pier 31, the acoustically clear zone, the compressibility was relatively high in approximately the upper 35 cm, but below this depth the compressibility was less than the precision of the measurements.

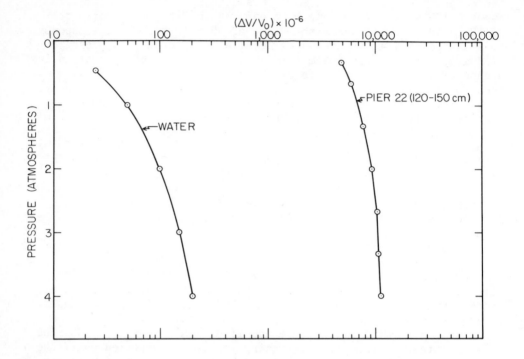

Fig. 8. The change in volume of a core segment at different pressures divided by the initial volume of the core segment, at atmospheric pressure, $\Delta V/V_0$, for a sample from 120-150 cm in a core
from pier 22. The same data for water are given for reference.
Analysis of a sample from 100-120 cm in a core from pier 31 showed
that $\Delta V/V_0$ was less than the sensitivity of the instruments, about
100×10^{-6}, at all pressures tested to 4 atm.

    It is apparent from Figures 8 and 9 that the sediment in the
acoustically turbid zone at pier 22 is much more compressible than
the acoustically clear sediment at pier 31. This greater compressibility is due to the presence of gas bubbles in the acoustically
turbid sediment, bubbles that are absent, or at least present in
very much smaller quantities at pier 31.

    Core samples from pier 28, an acoustically turbid zone, and
from pier 25, an acoustically clear zone, showed the same features
of compressibility as the samples from piers 22 and 31. The core
from 28 was more than two order of magnitude more compressible than
water, while the compressibility of the core from pier 25 could not
be distinguished from zero.

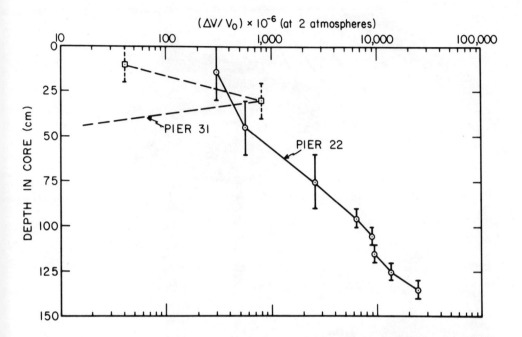

Fig. 9. $\Delta V/V_0$ at 2 atm for 1.5-meter cores from piers 22 and 31. The vertical lines represent the lengths of the core segments tested. The compressibility of the sediment was less than the sensitivity of the instrument below about 35 cm at pier 31.

The horizontal extent of the acoustically turbid sediments can be mapped from sub-bottom profiling records, but their vertical extent is not revealed by these records. In an attempt to determine the vertical extent of some of the acoustically turbid sediments, a two-legged probe was constructed (Figure 10). The legs, 6 meters long, are set 1 meter apart, and the lower end of each leg carries a 180 kHz transducer. One transducer serves as a source, the other as a receiver. The transducers are connected to the ship through cables. The source is driven by a signal oscillator, with the strength of the signal maintained at 5 volts. The strength of the received signal was measured in volts at a number of depths in the sediment. In use, the vessel is anchored bow and stern, and the weighted probe is dropped from the water surface (naviface) and allowed to fall freely into the soft sediment; penetration is normally 4 to 5 meters. Readings were made at the depth of maximum penetration, and usually at 0.25 meter intervals during recovery to the sediment surface. At each station, two to five 'drops' were made. An example of the agreement of the results typically observed on successive drops at the same location is shown in Figure 11. The

Fig. 10. Device used in determining the thickness of the acoustically turbid layers. The probe is dropped from the naviface and
allowed to fall freely into the sediment. The lower end of each
leg carries a 180 kHz transducer; one serves as a source, the other
as a receiver. The strength of the source is maintained at 5 volts,
and the strength of the signal received over a 1 meter path is measured. Measurements are made during recovery of the probe.

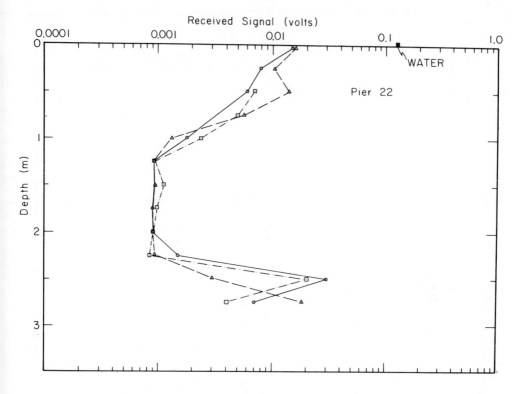

Fig. 11. Results of three successive 'drops' at the same location near pier 22 of the device shown in Figure 10. The strength of the signal in volts received from a 180 kHz transducer over a 1 meter path through the sediment is plotted against depth. The strength of the source signal was set at 5 volts. The strength of the signal over a 1 meter path through water is given for reference.

greatest uncertainty was in determining the actual depths of the probes below the sediment surface. The reported depths are probably not better than ±10 cm.

Readings from two different locations within the acoustically clear zone near pier 31, and from two locations within the acoustically turbid zone near pier 22 are presented in Figure 12. The strength of the received signal over a 1 meter path through water is given for reference. The data clearly show that in the acoustically clear zone the received signal was relatively strong and constant with depth. In the acoustically turbid zone near pier 22, the strength of the received signal was generally relatively strong in the upper 0.25-0.50 meter, but below this depth it decreased

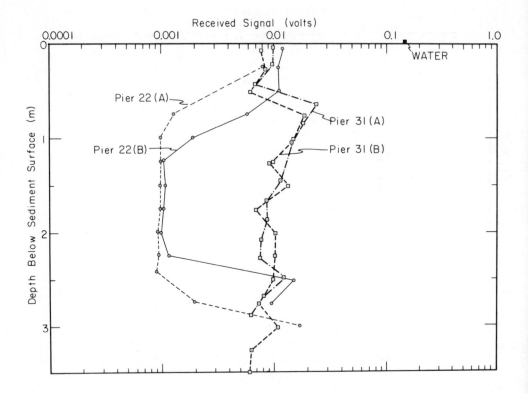

Fig. 12. Strength of signal in volts received from a 180 kHz transducer over a 1 meter path through sediment at two locations in an
acoustically turbid zone (pier 22), and at two locations in an
acoustically clear zone (pier 31). The strength of the source signal was set at 5 volts. Both the source and the receiver were
180 kHz transducers. The strength of the signal received over a
1 meter path through water is given for reference.

rapidly, reaching a minimum at a depth of about 0.75-1.25 meters.
Below this depth, the signal remained relatively constant and weak
to about 2.25-2.50 meters below the sediment surface, and then
increased over the next 0.5 meter. These results agree well with
the compressibility measurements shown in Figure 9. The data presented in Figure 12, as well as other data, show that in the acoustically turbid zone near pier 22, an acoustically turbid layer approximately 1 to 1.5 meters thick is sandwiched between a thin
acoustically clear surface layer and a lower acoustically clear layer. The vertical extent of this lower layer is uncertain, but other
data indicate that it is thin, probably less than 1 meter, and that
it is underlain by another layer of acoustically turbid sediment.

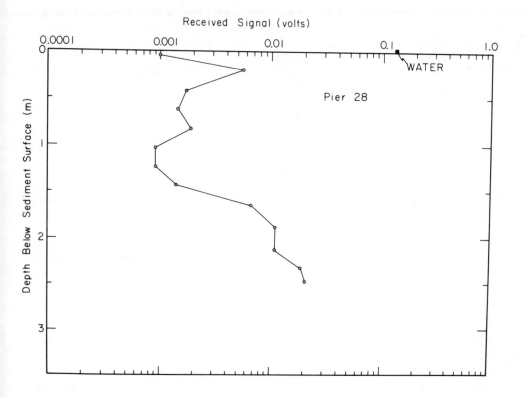

Fig. 13. Strength of signal received from a 180 kHz transducer over a 1 meter path through sediment in an acoustically turbid zone near pier 28. The strength of the source signal was set at 5 volts. Both the source and receiver were 180 kHz transducers. The strength of the signal over a 1 meter path through water is given for reference.

Results from another acoustically turbid area, pier 28, are shown in Figure 13. These data indicate that there was strong attenuation of the signal at this location throughout most of the upper 1.5 meters.

Figures 11 through 13 show that the acoustically turbid character of the sediments near piers 22 and 28 is produced by relatively thin (1-2 meters) layers. The data also show that at a frequency of 180 kHz the attenuation is so great that a 1 meter layer (the path length between transducers) can be considered acoustically to be infinitely thick. High-resolution sub-bottom profiling records, made with a sound source whose peak energy is at 5 kHz, show that even at this lower frequency the reflection from these layers is so great that the underlying features are masked.

Gravity cores were taken near pier 22, the acoustically turbid zone, and near pier 31, the zone of good acoustic penetration, to determine if there are any important differences in the particle size distribution, the water content, and the organic carbon content of the sediments in these two 'zones'. Size analyses, by standard pipette methods, failed to show any significant differences either with depth within a core or between cores; both were silty clay. The water contents (% wet mass) are shown in Figure 14. The sediment at pier 22 had slightly higher water content throughout most of the core than at pier 31, but the differences were relatively small. The amount of organic carbon (% dry mass) was considerably higher in most of the upper 0.5 meter in the acoustically turbid zone, but below 0.75 meter the difference was quite small (Figure 15). The organic carbon is the difference between the total carbon determined by combustion in an induction furnace and carbonate carbon determined gasometrically. It is possible that the concentrations of organic carbon were originally high throughout the sediment at pier 22, the organic matter decomposed below the upper 0.5 meter, and this decomposition produced the gas bubbles which cause reverberation of sound within the sediment. It is not clear why the organic matter should be higher in the upper 0.5 meter at pier 28 than at pier 22.

Not all acoustically turbid zones are, however, produced by the presence of gas bubbles. Analyses of cores from acoustically turbid zones in the Chester River estuary, a tributary of the Chesapeake Bay, have indicated that some acoustically turbid zones are produced by buried shell beds. This type of acoustically turbid zone can generally be identified by its morphology and its surface features. Buried shell beds are usually mound-shaped, and have much more irregular upper surfaces than do the acoustically turbid zones produced by gas bubbles.

CONCLUSIONS AND SOME SUGGESTIONS FOR FUTURE RESEARCH

The acoustically turbid sediments near the Chesapeake Bay Bridge at Annapolis have much higher compressibilities than acoustically clear sediments of similar grain size, mineralogy, and water content in the same region. The increased bulk compressibility of the acoustically turbid sediments is produced by the presence of gas bubbles entrapped in the interstices of the sediment. X-ray radiographs of cores have shown that gas bubbles are abundant in the acoustically turbid sediments, and scarce or absent in cores from adjacent acoustically clear areas. The acoustically turbid character of a bottom can be produced by relatively thin ($\leq$ 2 meters) layers of sediment that have high concentrations of gas bubbles.

The increase in bulk compressibility produced by the interstitial gas bubbles results in a decrease in the speed of sound in these

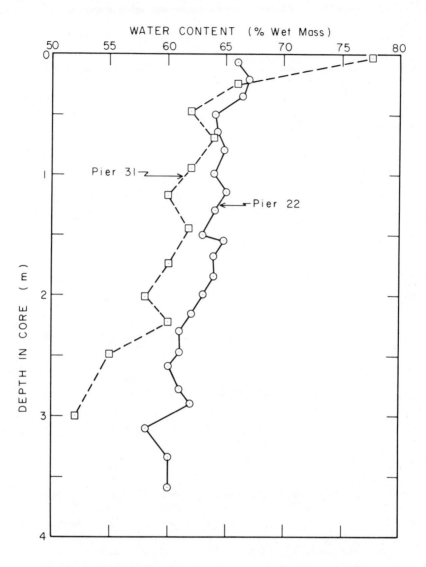

Fig. 14. Water content (% wet mass) versus depth for a core from the acoustically turbid zone near pier 22, and for a core from pier 31, an acoustically clear zone.

sediments, and an increase in their acoustical reflectivity. No measurements have been made of the speed of sound in the sediments near the Chesapeake Bay Bridge, but the data reported here indicate that the sound speed in the acoustically turbid sediments is probably very low. If *Wood's* [*1941*] equation for the speed of sound in

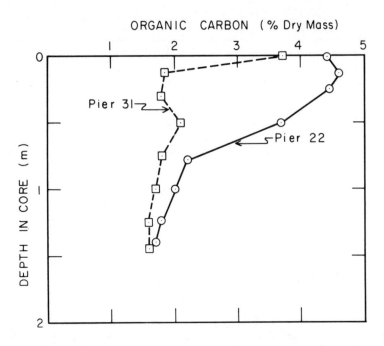

Fig. 15. Organic carbon (% dry mass) versus depth for a core from pier 22, an acoustically turbid zone, and for a core from pier 31, an acoustically clear zone.

gas-liquid mixtures can be applied to gas-mud-liquid mixtures, it indicates that if gas bubbles make up as little as 0.1% of the sample by volume, they will decrease the speed of sound to less than one-third the value with no bubbles. *Wood's [1941]* theory is valid for gas-liquid mixtures when the frequency of the sound is low compared to the resonant frequency of the largest bubbles present. The bubble content of the acoustically turbid sediments near the Chesapeake Bay Bridge may be as high as 1% of the material by volume. The sound speed in such a mixture might be as low as 100 m sec^{-1}, and it is unlikely that it would be higher than 250 m sec^{-1}. Sedimentary bottoms that have acoustic velocities as low as this behave as pressure-release surfaces, and have been reported in a number of areas [*Jones et al. 1958, 1964; Bobber, 1959; Levin, 1962*].

Determinations should be made of the sound velocity in the acoustically turbid and in the acoustically clear sediments in this area of Chesapeake Bay. Determinations should also be made of the size distribution that the entrapped gas bubbles have *in situ.*

In addition, research should be directed toward determining why such large variations in gas bubble content occur over such small distances, both vertically and horizontally.

ACKNOWLEDGMENTS

I thank G.M. Schmidt, W.B. Cronin, R. Moose, C.H. Morrow, and J.B. McKeon for their assistance both in the field and in the laboratory. The testing devices, which made this study possible, were designed by E.W. Schiemer.

This work was supported in part by the Oceanography Section, National Science Foundation, NSF Grants GA-21169 and GA-24786; and in part through a jointly funded project by The Fish and Wildlife Service, Bureau of Sport Fisheries and Wildlife, Department of the Interior through Dingell-Johnson Funds, Project F-21-1.

Contribution No. 188 from the Chesapeake Bay Institute of The Johns Hopkins University.

REFERENCES

Anderson, A. L., T. G. Muir, R. S. Adair, and W. H. Tolbert, A geoacoustic study of the Brazos River, Part I, *Environ. Studies, Defense Res. Lab. Tech. Rep. No. A-294* (DRL-A-294), Univ. of Texas at Austin, 1968.

Bobber, R. J., Acoustic characteristics of a Florida lake bottom, *J. Acoust. Soc. Amer., 31,* 250, 1959.

Emery, K. O., *A Coastal Pond Studied by Oceanographic Methods,* 80 pp., Elsevier Pub. Co., New York, 1969.

Grim, M. S., C. L. Drake, and J. R. Heirtzler, Sub-bottom study of Long Island Sound, *Geol. Soc. Amer., Bull., 81,* 649, 1970.

Hampton, L. D., Acoustic properties of sediments, *Defense Res. Lab. Acoustic Rep. 254* (DRL-A-254); Univ. of Texas, 1966.

Jones, J. L., C. B. Leslie, and L. E. Barton, Acoustic characteristics of a lake bottom, *J. Acoust. Soc. Amer., 30,* 142, 1958.

Jones, J. L., C. B. Leslie, and L. E. Barton, Acoustic characteristics of underwater bottoms, *J. Acoust. Soc. Amer., 36,* 154, 1964.

Klein, M. and H. E. Edgerton, Sonar--a modern technique for ocean exploration, *Inst. Elect. Electron. Eng. Spectrum, 5,* 40, 1968.

Levin, F. K., The seismic properties of Lake Maracaibo, *Geophysics 27,* 35, 1962.

Moody, D. W., and E. D. van Reenan, High-resolution subbottom seismic-profiles of the Delaware Estuary and Bay mouth, *U. S. Geol. Survey Prof. Paper 575-D*, D247, 1967.

Nyborg, W. L., I. Rudnick, and H. K. Schilling, Experiments of acoustic absorption in sand and soil, *J. Acoust. Soc. Amer. 22*, 422, 1950.

Reeburgh, W. S., Observations of gases in Chesapeake Bay sediments, *Limnol. and Oceanogr., 14*, 368, 1969.

Schubel, J. R., and E. W. Schiemer, The origin of acoustically turbid sediments in Chesapeake Bay, *Spec. Rep. 23, Ref. 72-5*, 21 pp., Chesapeake Bay Inst. of The Johns Hopkins University, 1972.

Schubel, J. R., and C. F. Zabawa, A Pleistocene Susquehanna River channel connects the lower reaches of the Chester, Miles, and Choptank Estuaries, *Spec. Rep. 24, Ref. 72-8*, 12 pp., Chesapeake Bay Inst. of The Johns Hopkins University, 1972.

Schubel, J. R., and E. W. Schiemer, The cause of the acoustically impenetrable, or turbid, character of Chesapeake Bay sediments, *Mar. Geophys. Res.*, in press, 1973.

Skempton, A. W., Effective stress in soils, concrete, and rocks, in *Intern. Soc. of Soil Mechanics and Found. Eng., British Nat. Soc., Pore Pressure and Suction in Soils*, Butterworths, London, 1961.

Stiles, N. T., Isopachous mapping of the lower Patuxent Estuary sediments by continuous seismic profiling techniques, *U. S. Naval Oceanogr. Office Informal Rep. No. 37*, 26 pp., 1970.

Stiles, N. T., L. R. Breslau, and M. D. Beeston, The riverbed roughness and sub-bottom structure of the main shipping channel to Sai Gon, RVN (Nga Bay, Long Tau, Nha Be, and Sai Gon Rivers), *Proc., 12th Conf. on the Naval Minefield, Naval Ordnance Lab. Tech. Rep. 69-69, 1*, 327, 1969.

Wood, A. B., *A Textbook of Sound*, 578 pp., MacMillan Co., New York, 1941.

Wood, A. B., and D. E. Weston, The propagation of sound in mud, *Acustica, 14*, 156, 1964.

IN SITU INDICATIONS OF GAS HYDRATE

George M. Bryan

Lamont-Doherty Geological Observatory
Columbia University
Palisades, New York   10964

ABSTRACT

Unusually high seismic velocities in gas-rich sediments of the Blake-Bahama outer ridge suggest the possibility of gas hydrate deposits in the upper few hundred meters. A prominent reflector lies at a depth that coincides with the lower limit of the hydrate zone, as calculated on the basis of reasonable thermal gradients. A similar reflector in the Bering Sea is associated with a more normal velocity profile and with the presence of gas only at the reflector itself.

INTRODUCTION

The possible existence of gas hydrates in marine sediments was suggested by *Stoll et al.* [*1971*] to explain unusually high seismic velocities in gas-rich sediments on the Blake-Bahama outer ridge [*Hollister et al. 1972*]. A gas hydrate is a crystalline structure of water molecules in which gas molecules are physically trapped. As in the case of ice, the crystalline structure results from hydrogen bonding between water molecules, but the geometry is quite different from the hexagonal ice structure. In particular, the hydrate structures contain cavities, each of which can enclose a single molecule of the gas phase. Chemical bonding between the gas molecules and the water structure is not involved, so that a number of gases, all with approximately the same size molecules, can form hydrates which are structurally the same [see *Hand et al.; Hitchon;* and *Miller, 1974*].

In the presence of gas molecules of the appropriate size, and

299

at pressures in an appropriate range, many hydrates can form at temperatures above the freezing point of water. The result is an icelike solid. Thus, if the interstitial water of a saturated sediment were converted to a gas hydrate, a substantial increase in wave velocity could be expected [Stoll, 1974].

It turns out that the conditions of temperature and pressure required for the formation of natural gas hydrates are met in the upper sedimentary layers over much of the ocean floor. If there is a sufficient supply of gas in such a region, the interstitial water should be hydrated. Since 1971, when the hypothesis was suggested, seismic results that are more or less similar to those on the Blake-Bahama outer ridge have been reported from several other areas, notably the Umnak plateau in the Bering Sea.

THE BLAKE-BAHAMA OUTER RIDGE

The Blake-Bahama outer ridge is a large topographic feature extending southeast from the Blake plateau off the coast of Florida (Figure 1). Deep seismic profiling (Figure 2) has shown that it is comprised of a vast accumulation of sediment lying on the horizontal sequence known as Horizon A [Ewing and Ewing, 1964]. Since then, speculation with regard to the formation of the ridge has been based for the most part on differential deposition by ocean currents. Ewing et al. [1966] suggested that much of the sediment involved may have been eroded from the Blake plateau by the Florida current (Gulf Stream). Heezen and Hollister [1964] and Heezen et al [1966] proposed that the ridge was built by the deep contour-following current which flows generally south along the western margin of the North Atlantic. In an attempt to explain in a somewhat more quantitative way the location, orientation, and shape of the ridge, Bryan [1970] invoked both these currents, suggesting that they interact to form a stable flow pattern which is consistent with the observed depositional pattern and therefore could have been responsible for the initiation and early growth of the ridge on a flat sea floor. In a detailed study of seismic reflection records, Markl et al. [1970] show how this simple picture is complicated in the later stages of formation, as geostrophic effects become more dominant. The similarity between the sedimentary structure of the ridge and that of the normal continental rise to the north is pointed out, and it is suggested that a normal continental rise would have been built against the steep Blake escarpment if it were not for the deflecting influence of the Florida current.

One of the features common to the ridge and the normal rise to the north is a prominent reflecting horizon which parallels the sea floor at a depth corresponding to about 0.6-0.7 second of two-way travel time. This reflector cuts across a series of closely spaced reflectors that appear to be bedding planes (Figure 3) and

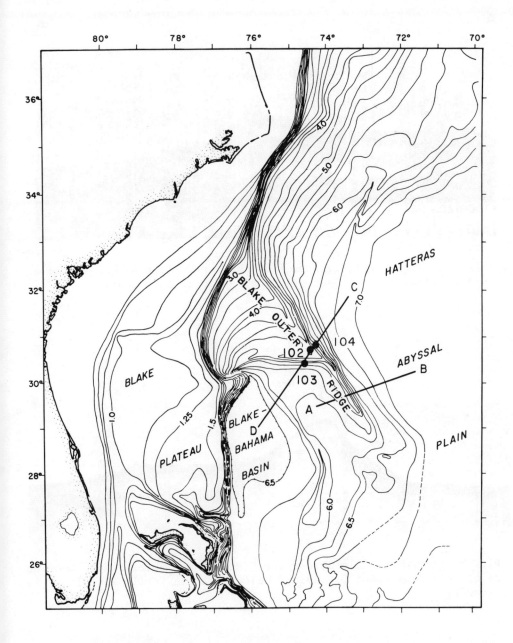

Fig. 1. Topographic contour chart of the Blake-Bahama outer ridge and environs. Lines AB and CD are profiler lines for Figures 2 and 3 respectively. Numbered points are DSDP drill sites.

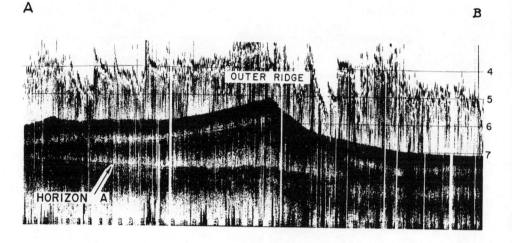

Fig. 2.   Early explosives profile showing Horizon A beneath the Blake-Bahama outer ridge.   Vertical scale is in seconds.

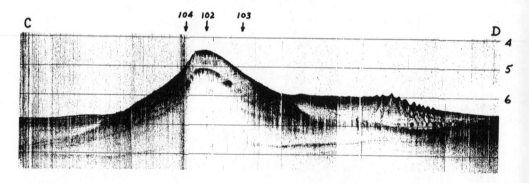

Fig. 3.   Airgun profile across the Blake-Bahama outer ridge, showing major reflector about 0.6 second below bottom.   Vertical scale is in seconds.

thus cannot itself be a stratigraphic boundary.   One of the major objectives of Leg 11 of the Deep Sea Drilling Project was to investigate the nature of this reflecting horizon.

Three holes on the Blake outer ridge were drilled on DSDP Leg 11 [*Hollister, Ewing et al. 1972*].   These sites are located on

Figures 1 and 3. Hole 102 was drilled on the crest of the ridge at
a water depth equal to an acoustic travel time of 4-1/2 seconds;
holes 103 and 104 on the southwest and northeast flanks respectively,
in a water depth equal to approximately 5 seconds travel time. The
cores from all three holes consisted largely of hemipelagic mud and
contained substantial quantities of gas. In many of the cores,
enough gas evolved to extrude sections of core from the plastic
liner, creating visible gaps. The gas was predominantly methane,
with traces of ethane. Age data show that rapid deposition characterized the building of the ridge, at least for the upper 600
meters. The accumulation rate was as high as 20 cm/1000 yrs in middle Miocene [*Ewing and Hollister, 1972*].

In holes 102 and 104, a major increase in resistance to drilling
occurred at about 620 meters below the sea floor. Correlation of
this drilling break with the prominent reflector described above implies a seismic velocity of 2.0 to 2.2 km/sec. (Hole 103 was abandoned before the reflector was reached and no velocity estimate was
possible.) Sound velocity in excess of 2 km/sec seems unusually
high for the hemipelagic mud encountered in these holes. However,
no alternative correlation is available; there is only one prominent
reflector and one major drilling break. If a lower velocity is assumed for the lithologically-resistant layer, there is nothing in
the drilling record to associate with the reflector, and no other
reflector to correspond to the drilling break. The change in drilling rate is quite sharp, amounting to an order of magnitude in less
than 20 meters, or about a wavelength at the seismic frequencies
involved. It is difficult to imagine a difference in hardness or
rigidity this large without a significant mismatch in acoustic impedance.

Corroboration of anomalously high sound velocity in the upper
sediments of the outer ridge comes from recent sonobuoy measurements
in the area. Preliminary results indicate an average velocity of
1.9 to 2.0 km/sec in the sediment above the prominent reflector.
As the sonobuoy method does not involve the drilling record, these
results constitute an independent check on the validity of the correlation made above.

The occurrence of such high velocities in hemipelagic mud, together with the evolution of large amounts of natural gas from that
mud, suggests the presence of gas hydrates in the sediment column
above the prominent reflector. That the thermodynamic conditions
are favorable here can be seen from Figure 4, which shows the thermodynamic equilibrium curve for methane hydrate on a temperature-
depth plot. Also shown are a set of curves giving temperature versus depth in the sediment for an assumed temperature of 3°C at the
sea floor depth (3300 meters in this case) and a range of assumed
thermal gradients from .02 to .08°C/m. For a gradient in the vicinity of .04°C/m, Figure 4 shows that the upper 500-600 meters of

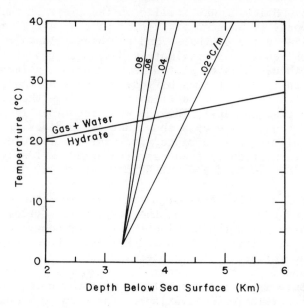

Fig. 4. Hydrate equilibrium curve and temperature versus depth for several thermal gradients, assuming a water depth of 3300 meters and a bottom temperature of 3°C.

sediment is in the hydrate region. Higher gradients predict correspondingly shallower hydrate zones.

The small slope of the hydrate equilibrium line in Figure 4 (approximately .002°C/m), compared with reasonable thermal gradients in the sediment column (approximately .04°C/m), results in very small changes in thickness of the predicted hydrate zone with changes in bottom depth. For example, at .04°C/m a 500 meter topographic change results in only a 25 meter change in thickness of the hydrate zone. Thus, the lower boundary of the hydrate zone would tend to parallel the sea floor at a depth of a few hundred meters below it.

It seems reasonable, therefore, to try to explain the prominent reflector in terms of the lower boundary of the postulated hydrate zone, particularly because the reflector appears to transgress bedding planes in the seismic records (Figure 3). This transgression is supported by the drilling results, which place the reflector at the Pliocene-Miocene boundary in hole 102, and no later than middle Miocene in holes 103 and 104 [*Ewing and Hollister, 1972*]. The reflector is probably associated with a diagenetic process, and its nearly uniform depth of about 600 meters below the sea floor matches

the hydrate model suggested above for a thermal gradient between
.03 and .04°C/m [see *Stoll, 1974* for further discussion of thermal
gradients]. The reflector might then be explained as resulting
from the velocity contrast between hydrated sediment above and normal (but possibly diagenetically altered) sediment below.

There are two major obstacles for this simple picture: first,
the drilling resistance increased at the interface, whereas a decrease seems more reasonable in going from hydrated to normal sediment; second, preliminary sonobuoy results indicate that the average
velocity in the 800 meters below the reflector does not decrease
but in fact may be as high as 2.2 km/sec.

Velocities in this range are about right for typical sediments
in this depth of burial. However, this value is very close to that
estimated for the sediment above the reflector, so that there appears to be insufficient velocity contrast across the reflector to
produce the recorded intensity. On the other hand, if gas is migrating upward from below and is impeded by hydrate above, it will
accumulate at the lower boundary of the hydrate zone as free gas,
and could cause a significant decrease in velocity in that region.
This is shown schematically in Figure 5. Velocity as a function of
burial for typical deep sea sediments is represented by the smooth
curve. The broken curve shows how this normal behavior might be
modified by the presence of hydrate in the upper sediments, and by
accumulation of free gas just below the hydrate zone. The narrow
low velocity band could cause the reflector in the profiler record,
but would not be resolved by the sonobuoy measurements.

This picture is admittedly rather speculative at the present
time. More detailed velocity profiles, as well as sampling with
pressure core samplers, are needed before a satisfactory explanation
is possible. However, there is perhaps some additional support for
the idea that free gas is involved at the reflecting horizon in the
fact that the reflector is most prominent in the vicinity of the
ridge crest and tends to fade out on the flanks [*Markl et al. 1970*].
This might indicate an accumulation of gas at the crest of the hydrate boundary. The true dips involved are of the order of a few
degrees.

THE BERING SEA

A reflector similar to the one on the Blake Ridge is recorded
in the Bering Sea [*Scholl et al. 1968*]. It appears about 0.8 second below the sea floor and tends to parallel the sea floor, cutting across bedding planes in order to do so. JOIDES hole 185 was
drilled just east of the Umnak plateau in the southeastern corner
of the Bering Sea. The area was apparently subjected to relatively
rapid deposition of terrigenous sediment. At this site, gas was

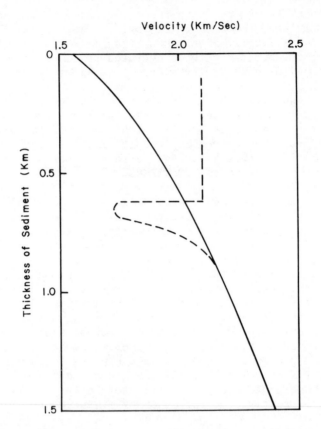

Fig. 5. Typical velocity profile (solid line) and hypothetical modification of it (dashed line) by the presence of hydrate in the upper 600 meters and accumulation of gas at 600 meters.

recovered at a depth of about 670 meters, which was taken to be the depth of the reflector, but no gas was found above this depth in noticeable quantities. Moreover, the average velocity computed from reflection time and drill-string depth was 1.65 km/sec, a typical value for the upper few hundred meters. Shipboard measurements on the cores gave a range of 1.5 to 1.75 km/sec, in good agreement with the average *in situ* value. Below 670 meters, shipboard measurements yielded 2.0 km/sec, and the transition from 1.75 to 2.0 km/sec occurs over about 20 meters.

Thus, from the point of view of the velocity profile, the Umnak plateau case is not as anomalous as the Blake Ridge. The velocity is reasonable for unconsolidated sediment down to about 670 meters, there is a sharp transition at this depth with high enough velocity contrast to account for the reflector, and the velocity below this depth is typical of the deeper sediments. On the basis of these facts alone, there is no justification for invoking the gas hydrate hypothesis, particularly as no gas was evident in the upper sediments. Aspects of this case, which are anomalous and which this case has in common with the Blake Ridge, are the presence of gas at the reflector depth and the transgressive nature of the reflector. It is this similarity that makes it tempting to assign a common cause to both cases. In the absence of further evidence, however, this can only be speculation.

The ultimate evidence, of course, would be a sample of sediment, recovered under *in situ* pressure and temperature that could be examined for the presence of gas hydrate. Preliminary attempts to obtain a sample in a pressure core barrel were made at hole 185, but were unsuccessful because of difficulties with the sampling device. The sampler is presently being redesigned for use at future sites where the presence of hydrate is indicated.

SUMMARY

Unusually high seismic velocities in gas rich sediments constitute evidence for the existence of gas hydrate in the upper part of the sedimentary column on the crest of the Blake-Bahama outer ridge. Temperature and pressure conditions are appropriate for the formation of hydrate through the upper few hundred meters of sediment. A reasonable estimate of temperature gradient places the lower limit of the hydrate zone in the vicinity of the major reflector which parallels the sea floor and cuts across stratigraphic isochrons. The reflector might be the result of a drop in velocity at this boundary from a value associated with a hydrated sediment to one associated with a gassy sediment.

A similar reflector in the Bering Sea is associated with a more normal velocity profile and with the presence of gas only at the reflector itself.

ACKNOWLEDGMENTS

This paper is based in part on research sponsored by the National Science Foundation, under Grant GA 30881 to Columbia University, and by the Office of Naval Research, under Contract N00014-67-A-0108-0004.

Lamont-Doherty Geological Observatory Contribution No. 2021.

REFERENCES

Bryan, G. M., Hydrodynamic model of the Blake Outer Ridge, *J. Phys. Res.*, *75*(24), 4530, 1970.

Ewing, J., and M. Ewing, Distribution of oceanic sediments, in *Studies in Oceanography*, p. 525, (Geophys. Inst. Univ. Tokyo), Tokyo, 1964.

Ewing, J., M. Ewing, and R. Leyden, Seismic profiler survey of the Blake Plateau, *Bull. Amer. Ass. Petrol. Geolog.*, *50*(9), 1948, 1966.

Ewing, J., and C. H. Hollister, *Initial Reports of the Deep Sea Drilling Project*, vol. 11, pp. 951-973, U. S. Government Printing Office, Washington, 1972.

Hand, J. H., D. L. Katz, and V. K. Verma, Review of gas hydrates with implications for ocean sediments, in *Natural Gases in Marine Sediments*, edited by I. R. Kaplan, pp. 179-194, Plenum Press, New York, 1974.

Heezen, B. C., and C. D. Hollister, Deep-sea current evidence from abyssal sediments, *Mar. Geol.*, *1*, 141, 1964.

Heezen, B. C., C. D. Hollister, and W. F. Ruddiman, Shaping of the continental rise by deep geostrophic contour currents, *Science*, *152*(3721), 502, 1966.

Hitchon, B., Occurrence of natural gas hydrates in sedimentary basins, in *Natural Gases in Marine Sediments*, edited by I. R. Kaplan, pp. 195-225, Plenum Press, New York, 1974.

Hollister, C. D., J. I. Ewing, *et al.*, *Initial Reports of the Deep Sea Drilling Project*, vol. 11, pp. 9-50, U. S. Government Printing Office, Washington, 1972.

Markl, R. G., B. M. Bryan, and J. I. Ewing, Structure of the Blake-Bahama Outer Ridge, *J. Geophys. Res.*, *75*(24), 4539, 1970.

Miller, S. L., The nature and occurrence of clathrate hydrates, in *Natural Gases in Marine Sediments*, edited by I. R. Kaplan, pp. 151-177, Plenum Press, New York, 1974.

Scholl, D. W., E. C. Buffington, and D. M. Hopkins, Geologic history of the continental margin of North America in the Bering Sea, *Mar. Geol.*, *6*, 297, 1968.

Stoll, R. D., Effects of gas hydrates in sediments, in *Natural Gases in Marine Sediments*, edited by I. R. Kaplan, pp. 235-248, Plenum Press, New York, 1974.

Stoll, R. D., J. Ewing, and G. Bryan, Anomalous wave velocities in sediments containing gas hydrates, *J. Geophys. Res.*, *76*(8), 2090, 1971.

PAGODA STRUCTURES IN MARINE SEDIMENTS

K.O. Emery

Department of Geology and Geophysics
Woods Hole Oceanographic Institution
Woods Hole, Massachusetts   02543

ABSTRACT

Geophysical traverses across flat or nearly flat mud bottom at depths between 2000 and 5500 meters off western Africa permitted the making of extensive shallow-penetration recordings at 3.5 kHz. The recordings reveal the common presence of alternating dark and light triangular features whose internal structure and acoustic properties may be due to local centers of cementation by gas hydrates, or clathrates. Probably the features are widely distributed in the fine-grained sediments of continental rises and abyssal plains of the world ocean, but are not detected by the usual echo sounding at 12 kHz or seismic reflection profiling at 20 to 100 Hz.

INTRODUCTION

A unique structure in fine-grained layered sediments of Lake Geneva was described by *Glangeaud et al.* [*1964*] and by *Serruya et al.* [*1966*] from echo sounding profiles. It consists of a long series of triangles (almost certainly cones if the third dimension were known), with triangles having their apex pointing upward, alternating with ones pointing downward. Those pointing downward have a convex upward base and are light gray in the recordings owing to low acoustic energy reflected from internal layers. The alternate triangles that point upward are dark gray because more acoustic energy is reflected from internal layers. The layers themselves can be traced throughout long series of the triangles, and they are convex upward in the light-gray triangles and generally convex downward in the dark-gray ones. The dark-gray triangles have steep curved side boundaries reminiscent (in the recordings) of the roofs of oriental pagodas; thus *Serruya et al.* termed the features pagoda structures.

Shipboard investigations of the Eastern Atlantic Continental Margin program of the International Decade of Ocean Exploration during 1972-73 revealed a widespread distribution of these pagoda structures on the deep floor of the Atlantic Ocean off western Africa (Figure 1). Altogether *R/V ATLANTIS II* made about 60,000 line km of geophysical traverses in the area of Figure 1. Well developed pagoda structures were noted along 5950 km and less clear ones along 7150 km. Thus they were observed along about 22% of the total traverse length. Most of the traverses that contained no pagoda structures were in areas of thin or no sediment (the rocky Mid-Atlantic Ridge, lateral ridges, and seamounts); others were along the continental shelf where perhaps the cause was somewhat coarser-grained sediment. Absence along some traverses across some of the fine-grained deep-water sediment may have been due to improper adjustment of the recorder (such as too low a gain or too fast a sweep rate), but most of the absence is attributed to lack of some as yet unknown properties of the sediment that are required for pagoda structures to form.

Probably the pagoda structures are common in the entire world ocean, but have not been reported by marine geologists because most acoustic sounding is done at too high or too low acoustic frequencies, and recordings may be made at too fast a sweep rate. The observations are reported here as other investigators may be interested in questions of the distribution and possible origin of the structures.

DESCRIPTION

The light-gray triangles (apex down, base convex upward) of the ocean-floor pagoda structures are as long as 5 km, but more typically are 200 to 500 meters long. Their convexities rise as high as 4 meters above their perimeters, though some are barely perceptible. The total height or thickness ranges up to 50 meters (Figure 2). Minimum lateral dimensions probably are meaningless, as examples appear to exist to below the limit of resolution through several km of water, even though the signals are only 2 to 4 meters apart along the ship traverses (one-second repetition rate). Side slopes of the convexity above the light-gray triangles attain about one degree, according to the recordings. These dimensions contrast with the 10 to 15 meters height (thickness) and about 20 meters length of the structures in Lake Geneva that occur in water depths of only 30 to perhaps 160 meters.

The dark-gray triangles (apex upward) typically have about the same length bases as the light-gray ones, especially when the apex is near the sediment surface (Figure 2B, C). In some areas having sediments not obviously different in other respects, the apex of the dark-gray triangles appears to lie several meters beneath the

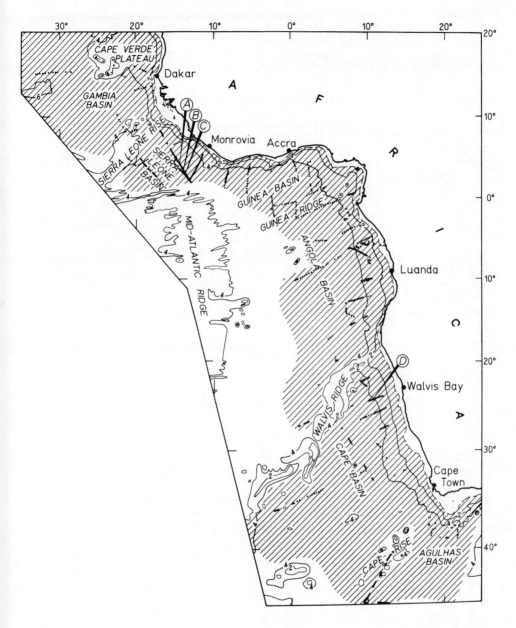

Fig. 1. Distribution of traverses of *R/V ATLANTIS II* along which pagoda structures were observed during cruises of 1972-73 off western Africa. Continuous lines denote the best records, dotted lines the poorer ones. The structures are confined to mud bottom, larger regions of which are denoted by the diagonally lined pattern beyond the shelf break. Letters indicate positions of recordings presented in Figure 2.

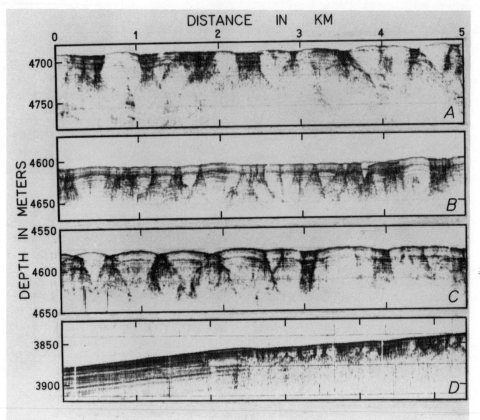

Fig. 2. Typical variations of the pagoda structures shown by photographs of the original 3.5 kHz recordings. (A) Hourglass (left) and irregular form of triangles (right), 7 Feb. 1972 (0527 to 0542), vertical exaggeration is 9.6; (B) Typical pagodas (left) and irregular ones (right), 7 Feb. 1972 (1058 to 1113), vertical exaggeration is 9.6; (C) Typical pagodas, 7 Feb. 1972 (1254 to 1310), vertical exaggeration is 9.6; (D) Hourglass to irregular pagodas (left) and undisturbed sediment (right), 14 April 1972 (0044 to 0117), vertical exaggeration is 10.5. Positions are shown by A, B, C, D on Figure 1.

sediment surface, but the boundaries continue upward so that the dark-gray areas are somewhat hourglass shaped (left part of Figure 2A). In still other areas, some of the features are more irregular than triangular (Figure 2A, D). Layers within the sediment continue unbroken through long series of alternating light-gray and dark-gray triangles.

Rather different results are obtained by sensors of different acoustic frequencies. The usual echo sounder at 12 kHz is designed to depict the surface of the ocean floor, and it fails to obtain subsurface reflections from within most sediments; thus it does not show pagoda structures. Acoustic energy from seismic profilers at 20 to 100 Hz penetrates deeply into the bottom, but the wave lengths of 15 to 75 meters are too long to detect thin layers, and their records are made at usually too fast a sweep rate to be useful. Intermediate is the 3.5 kHz unit that is designed especially for shallow penetration of muddy sediments and can obtain reflections from as deep as 100 meters in clay and silt. At a one-second sweep rate and using 49-cm wide paper, a reflector 40 meters below the bottom is drawn on the paper 2.6 cm below the bottom trace. Even with the 3.5 kHz system, the pagodas can be missed if the sweep rate is too slow (two-second versus one-second, for example) or if violent pitching of the ship produces gaps in the record through absorption of acoustic energy by air bubbles under the hull-mounted transducer.

In all examples the pagoda structures are restricted to the top layers of stratified sediment, never beneath undisturbed sediments. Water depths, where they were observed off western Africa, range from 2000 to 5500 meters. Pagodas occur mainly in flat-lying sediments, but also on slopes as steep as 0.5 degree. They are locally present in broadly folded sediments, but whether they formed before or after folding is unknown. They are not universally present even in flat-lying sediments; some examples (Figure 2, D) exhibit a sharp boundary between sediments containing pagoda structure and otherwise similar sediments having none of these structures. The pagoda structures are completely unlike the structures of continental-rise hills, and the distributions of pagodas and continental-rise hills also are different. Thus the two kinds of features are not easily confused.

The pagoda structures as shown on the usual recording (Figure 2) have exaggerated side slopes owing to the ten to fifteen times larger vertical than horizontal scales. When plotted at the same vertical as horizontal scale (Figure 3), the triangles are seen to be very low and broad, and the boundaries between the light-gray and the dark-gray triangles are very gentle, only 5-15 degrees. Thus, on true scale the pagodas become almost flat roofs. This true-scale version obviously is important for the assignment of origin to the structures.

ORIGIN

The pagoda structures appear to not be caused by instrumental artifacts such as acoustic interference. The 3.5 kHz signal has most of its energy in a wave length of 40 cm (in surface sediments), which is small in comparison with the size and thickness of the

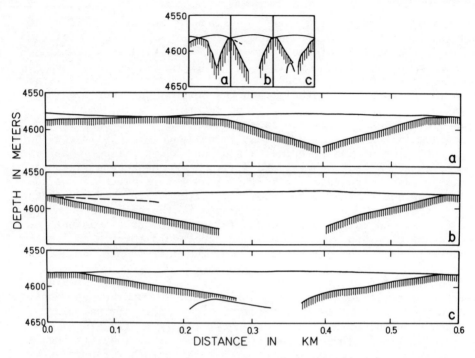

Fig. 3. Examples of pagoda structures reproduced at identical vertical and horizontal scales show the relative flatness of the triangles. The three long panels are expanded versions of the three
sections of the top panel that in turn is a direct tracing of boundaries of pagoda structures in the photograph at the left part of
Figure 2C. Vertical hatching denotes the 'roofs' of the pagodas or
dark triangle.

features. Moreover, the widespread distribution of the structures,
their great range in water depth, their locally sharp limits, and
the clarity of the boundaries between the light-gray and dark-gray
triangles appear to be sufficient to eliminate the possibility of
their being caused by some kind of refraction or diffraction in the
overlying water or in the sediments. Recordings over irregular
rock bottom at any water depth are characterized by smoothly curving
hyperbolae produced by acoustic returns from small reflection surfaces. However, the boundaries between the light-gray and the dark-
gray triangles in layered muds generally are too irregular in shape
to be confused with such hyperbolae. Moreover, the junction of
these boundaries with the edges of the convex upward base of the
light-gray triangles is nearly everywhere angular, and not smoothly
curving as they would be if the boundaries were depthward continua-
tions of the apex of an hyperbola. Evidently, the origin of the
pagoda structures is geological, not instrumental.

Glangeaud et al. [*1964*] originally assigned the pagoda structures to periglacial frost wedging, but later *Serruya et al.* [*1966*] eliminated that origin because of the much greater thickness of the features within the sediments of Lake Geneva than in soils and surface sediments of other areas of Europe that had been much more strongly affected by glaciation. More compelling arguments against glacial origin are the much more gentle dips of the boundaries than those of frost-wedge cracks, and now of course the finding of the same features deep beneath the tropical Atlantic Ocean.

The alternation of light-gray and dark-gray triangles in layered surface sediments of the ocean floor implies patterns of alternating differences in acoustic properties of the bottom along the ship traverses. The light-gray triangles are areas where the contrast in acoustic impedance between different layers is less than between the same layers in the dark-gray triangles. Such differences in acoustic contrast can scarcely be original (syngenetic) features of the sediment. In addition, the upward convexity of both bases and layers of sediment within the light-gray triangles appear most unlikely to be original depositional features. If they had been initiated by an original irregular topographic surface, the next layer of sediments to be deposited would generally be expected to be concentrated in the deeper spots, thus tending to smooth the bottom instead of uniformly mantling it. The absence of smoothing layers, and for that matter, of buried (older) pagoda structures, implies that the structures are quite recent in age, affecting sediment previously deposited. It is also evident that the change in properties either has not occurred during previous episodes, or the change has reverted during the burial of previous pagoda structures.

The suggestion has been made by George M. Bryan of the Lamont-Doherty Geological Observatory (personal communication, 21 September 1973) that the features may be due simply to the geometry of stacks of hyperbolae caused by curved or irregular reflecting surfaces on and in the bottom. Such a suggestion requires the presence of large tracts of bumpy topography, but abyssal plains are notoriously flat, and the widespread distribution of suitably bumpy topography or internal structure is no less easy to explain through either patterns of sedimentation or subsequent erosion of the bottom.

One might rationalize that the upward convexity and the small acoustic contrast of layers within the light-gray triangles may be the result of local centers of cementation that are separated by areas of ordinary uncemented sediments. A mechanism worthy of consideration is the development of gas hydrates, or clathrates in surface marine sediments. Clathrates are gases in a crystalline ice-like solid state caused by high pressure and low temperature. They may exist in marine sediments, according to pressure-temperature theory, and to rather scant direct observation of sediment

properties [*Stoll et al. 1971*]. The hydrostatic pressure of 200
to 550 atm and the bottom temperature of less than $3°C$ on the sea
floor off Africa (Figure 1) are compatible with data points on experimental equilibrium curves of methane hydrate formation.

If clathrates do form and do materially reduce the permeability
of the sediments, they probably would block the upward movement of
gas (either in gaseous or dissolved form) derived from underlying
sediments. Thus, the initial formation of clathrates at many scattered points would result in gradually enlarging cones of cemented
sediments. Eventually the edges of adjacent growing cones must intersect to form roughly hexagonal patterns akin to those of mud
cracks, basalt columns, and confined soap bubbles. The upward convexity of the light-gray triangles suggests that the clathrates
acted as a cementing agent, preventing as much compaction of the
sediments as in the adjacent dark-gray low-clathrate triangles.
The downward pointing apex of the cones may have resulted from the
greater accumulation of gas beneath the centers of the cemented
areas than near their perimeters. Thus the steepness of a cone can
be a measure of the permeability of the sediment to movement of gas
versus the hydration rate of the gas. The clathrates may have lost
their cementing power (due to geothermal gradient?) by the time
deep burial occurs, or have diffused into the overlying sediments
and water so that the pagoda structures are flattened and lost to
the geological record.

Presumably, the causative gas is methane or carbon dioxide,
both of which are produced by decomposition of organic matter. As
organic matter is most abundant in clays and silts, especially at
times of high productivity, perhaps the present abundance of pagoda
structures is the result of high productivity of marine waters during the times of rapid circulation and high runoff associated with
Pleistocene glacial stages.

One other structure that has been ascribed to cementation by
clathrates is the thick reflector Y that underlies and parallels
the surface of the Blake-Bahama outer ridge, but through which
closely spaced reflectors due to bedding can be seen to cross
[*Markl et al. 1970; Ewing and Hollister, 1972; Lancelot and Ewing,
1972*]. Considerable quantities of methane were present in the drill
cores from the zone of the reflector Y.

Whatever their origin, the pagoda structures warrant further
investigation, both by surface ship and by research submersibles,
which offer the only possibility of direct observation. Whether
methane is associated with them remains to be established.

ACKNOWLEDGMENTS

Contribution No. 3171 of the Woods Hole Oceanographic Institution. These results are part of those from the Eastern Atlantic Continental Margin program of the International Decade of Ocean Exploration (Grant GX-28193 of the National Science Foundation). Appreciation is due the many watch standers and their assistants who carefully marked the records during the 12 months of cruising off western Africa.

REFERENCES

Ewing, J. I., and C. H. Hollister, *Initial Reports of the Deep Sea Drilling Project,* vol. 11, pp. 951-973, U. S. Government Printing Office, Washington, 1972.

Glangeaud, L., O. Leenhardt, and C. Serruya, Structures enregistrées par le "mud-penetrator" dans les sédiments quaternaires du Léman, *C. R. Acad. Sci., Paris, 258,* 4816, 1964.

Lancelot, Y., and J. I. Ewing, *Initial Reports of the Deep Sea Drilling Project,* vol. 11, pp. 791-799, U. S. Government Printing Office, Washington, 1972.

Markl, R. G., G. M. Bryan, and J. I. Ewing, Structure of the Blake-Bahama Outer Ridge, *J. Geophys. Res., 75,* 4539, 1970.

Serruya, C., O. Leenhardt, and A. Lombard, Etudes géophysiques dans le Lac Léman, Interprétation géologique, *C. R. Seances Soc. Phys. Hist. Natur. Genève, 19,* 179, 1966.

Stoll, R. D., John Ewing, and G. M. Bryan, Anomalous wave velocities in sediments containing gas hydrates, *J. Geophys. Res., 76,* 2090, 1971.

LIST OF CONTRIBUTORS

Aubrey L. Anderson, Applied Research Laboratories, The University of Texas at Austin, Austin, Texas 78712.

P.E. Baker, Chevron Oil Field Research Company, P.O. Box 446, La Habra, California 90631.

George M. Bryan, Lamont-Doherty Geological Observatory, Columbia University, Palisades, New York 10964.

George E. Claypool, Branch of Oil and Gas Resources, U.S. Geological Survey, Denver, Colorado 90225.

K.O. Emery, Department of Geology and Geophysics, Woods Hole Oceanographic Institution, Woods Hole, Massachusetts 02543.

Douglas E. Hammond, Lamont-Doherty Geological Observatory, Columbia University, Palisades, New York 10964.

Loyd D. Hampton, Applied Research Laboratories, The University of Texas at Austin, Austin, Texas 78712.

J.H. Hand, Department of Chemical Engineering, The University of Michigan, Ann Arbor, Michigan 48104.

David T. Heggie, Institute of Marine Science, University of Alaska, Fairbanks, Alaska 99701.

Brian Hitchon, Research Council of Alberta, Edmonton, Alberta, Canada T6G 2C2.

I.R. Kaplan, Institute of Geophysics and Planetary Physics, University of California at Los Angeles, Los Angeles, California 90024.

D.L. Katz, Department of Chemical Engineering, The University of Michigan, Ann Arbor, Michigan 48104.

Graeme L. Lyon, Institute of Nuclear Sciences, Department of Scientific and Industrial Research, Lower Hutt, New Zealand.

Richard D. McIver, Esso Production Research Company, Houston, Texas 77001.

Byron J. Mechalas, Southern California Edison Company, Rosemead, California 90053.

Stanley L. Miller, Department of Chemistry, University of California at San Diego, La Jolla, California 92037.

William S. Reeburgh, Institute of Marine Science, University of Alaska, Fairbanks, Alaska 99701.

J.R. Schubel, Chesapeake Bay Institute, The Johns Hopkins University, Baltimore, Maryland 21218.

Robert D. Stoll, School of Engineering and Applied Science, Columbia University, New York, New York 10027.

V.K. Verma, Department of Chemical Engineering, The University of Michigan, Ann Arbor, Michigan 48104.

Thomas Whelan, III, Coastal Studies Institute, Louisiana State University, Baton Rouge, Louisiana 70803.

SUBJECT INDEX